WHY NOT?

A Handbook
For Inventors And Entrepreneurs

By

ALAN E. THOMPSON

ISBN-13: 978-1517777173
ISBN-10: 1517777178

DEDICATION

For Alison, who sorted out my glitches!

CONTENTS

SECTION 1

PREFACE

This book is about ideas, how to generate them, how to evaluate them and how to develop and refine them. I hope to show how you may think outside the box without too much reference to the way things are supposed to be done. The basic approach is to unbolt thinking from the constraints of reality and what other people have done, then try to see if it can be bolted back on again, hopefully with a new viewpoint or way of looking at things.

While thinking in the abstract is the primary goal, thinking about concrete things like cars, ships and railways, plus a host of other machines and mechanisms, will lead to modifications that are better classed as inventions. Should you decide to do a bit of inventing, or are already an inventor, an outline of the patent system will be given, plus pitfalls and problems that beset the inventor. Some real inventions will be disclosed within, 'by way of example', as the patent lawyer says.

Herein lies a problem. If you write a book, you have copyright as soon as it is written, or effectively on publication, which may be merely showing it to a friend or printing it in a book or magazine. Your publisher will lodge a few copies of your book with the British Library, or Library of Congress to make assurance doubly sure. Your copyright will also last for fifty years after your death, for your heirs and successors.

The poor inventor is in a much more parlous condition; he has to keep everything secret until he has a patent, which can be very expensive, and then he has only a limited time, typically fifteen years, in which to develop and market it. Even with a patent he may still have to take it through the courts if it is infringed, for a patent confers no right other than the right to defend, and again this will lead to great delay and expense.

The system is loaded against the small inventor, and can be difficult for a large corporation. The reason of course is that manufacturing companies have a large amount of money tied up in production facilities, and they do not welcome new inventions that may make their plant and machinery obsolete. Even when they are forced to yield, it is only slowly. The development of the Internet may well speed up the rate of change, for another problem with invention is the very long lead-time before things get accepted. This is probably due to the fact that once established in a position, people do not change their ideas very much. It is left to the next generation to make the world anew, after the old guard has retired or died off. Again, we are told how the speed of processors doubles every so many years, making the current year's computer obsolete before it is off the shelf. Actually, computer speeds have remained remarkably constant; as soon as capacity increases, it is filled with a hundred programs that very few people will ever use, or even know are there, that clog up the works and slow things down again. Also there is the problem of security. For the average person, not using a very secure system, it is best to assume that anything you say on the net or by cell-phone is effectively broadcast to all and sundry.

To try to circumvent this block to invention, the new devices and ideas in this book that have an undoubted commercial value are presented as a form of 'Disclosure in Confidence'. By this I mean that they are presented as an opportunity for a developer not to trick, by stealing them, but to treat, by way of negotiating an agreement. To sweeten the pill, royalty payments will be exceedingly low: think of fractional percentages, if you don't mind a mixed metaphor. A painless way might be to exchange the idea for some shares if a limited company is formed. Any income will go to development of ideas and hopefully some funding for good ones. The Institute of Patentees and Inventors can help; you should think of joining if you become an inventor.

This of course is a very weak form of protection, and will be largely based on goodwill and a marginal cost to the developer. However, there is a slight sting in the tail. We could of course make quite sure that you don't get a valid patent if you decide not to play ball.

One argument against this is the concept of, 'no contract without a consideration', however someone will have paid for the book, even if it has been given or loaned to the reader, so effectively the payment is a consideration. Quite often inventions are multiply submitted, particularly where there are a number of potential customers and no expensive machinery required, so the fact that more than one person will read the same information does not invalidate the potential contract. I might even sell a hundred books!

The main game however is thinking. Anybody can do it, and be as outrageous as they like. We will examine some oddball inventions, like a way to send information faster than the speed of light. I can show you an outline of a way in might be possible. Do you want to design a perpetual motion machine? The universe may in fact be one, and we can isolate elements of this and see how they could be applied! Do you want to send a letter back into the past? I can tell you the outline of the necessary requirements you will need to do so.

Design an anti-gravity device? Personally, I use a table, but there is an outside chance we may be able to design a 'levitator'; again, I will tell you how it might be done.

Ideas of this nature are often regarded as 'eccentric', and the term is usually regarded as derisory. However, the eccentric is a most important member of society; to start with, he defines the centre. Also, 'centrists' are in fact in a mutually exclusive position and have nowhere to go if their ambition is achieved. So be eccentric if it suits you; it is the only route to progress!

The chances of any of these oddball inventions being achieved at present look small. I tend to think of such ideas as wild geese, flying high and fast, free and unobtainable, diminishing towards their far horizon with perhaps the echo of a derisory honk, but in exploring such arcane and unpopular by-lines, other knowledge might be gained, and other items might be developed. After all, it is said that the quest for a non-stick frying pan led to men landing on the moon. (I may have got that bit the wrong way round.) Above all, it should be fun! Most readers will not be putting a professional career at stake by pursuing such ideas. The art is not to believe in them, but to keep an open mind, to suspend disbelief; above all they are tooth cutting tools and very puzzling and exciting ones at that.

For the person with a more practical turn of mind, there is plenty else. Trains that never slow or stop, but still allow people to get on and off, or engines that use a third less fuel than current designs. Also there are small and useful gadgets for around the home. A competent inventor can look at an industry, and make a very shrewd guess about

4

what they are working on. A good inventor can also predict what they will be working on next, for there is an overlap between invention and futurism. A very good inventor-cum-futurist will also know what they ought to be working on, but are not!

The book is a collection of small pieces, written over a considerable span of time for different audiences and markets. I have decided to leave it as such, apart from removing much of the duplication. I hope the reader will be able to adjust from a whimsical to a more serious style; to excuse the outright speculation, mixed in with solid science; the fully baked alongside the half-baked, and distinguish between the two. There are bound to be different bits that are of little interest to different people, and some may not be easily understandable. This is probably because I did not understand the subject too well myself, but such passages can be skipped. The first section deals with largely practical matters, with the more hypothetical items, which may be of greater interest to some, later in the book.

Regretfully, some health issues have led to a rather hurried final tidying-up, resulting in a few minor errors and duplications being left in, and diagrams and graphs not being included. Please accept my apologies for this.

INTRODUCTION

MY GEDANKEN DRAWER

Gedanken is a German word meaning 'thought'. A Gedanken Experiment, popularised by Einstein although not perhaps originated by him, is an experiment that you do in your head. This is necessary in some fields such as astronomy, as one cannot do experiments with stars in the same way as you can with a test tube of chemicals. It is also necessary in other fields, such as nuclear physics, where although you can do the experiment in reality, the apparatus is hugely expensive.

As an inventor and a futurist I tend to do quite a lot of Gedanken experiments, mainly because they are quicker and cheaper than the real thing. To prevent them getting mixed up with reality I have a 'Gedanken drawer', which of course is also imaginary, to keep them in. One interesting feature of this is that all kinds of ideas that would normally be categorised and separated get mixed up together, with sometimes surprising results, a confluence and fusion from different directions and disciplines. I had thought of coining the word confusion for this process, but decided to leave it, in case it was classified with cold fusion or otherwise misinterpreted.

This leads to a problem called the 'reality check', for you can do all kinds of experiments in thought only, without moving a finger, and get all kinds of results. However, there is the problem as to whether your results have anything to do with the real world.

An experiment, Gedanken or real, should be directed. It is really a sort of question asked to examine a problem or topic of interest, and the answer should relate to some other aspect of the initial problem. In other words, there can be a sort of internal consistency, which can act as a reality check: if you can ask the right questions, you are likely to get the right answers. The real problem is in asking the right questions.

Probably most real world experiments start off as Gedanken experiments, for you have to formulate an idea in your head before you take up a pencil, or a hammer. I say 'most', because on occasion the real world intrudes and does some variant of the experiment for you. One example is the fortuitous arrival of penicillium spores on Professor Flemings' petri dishes. He still had to take advantage of this however; a lot of people might have thrown them away and started again without the accidental penicillium mould. Thus it is important to be aware that you might get results other than those you originally expected, and this might tell you more than if you had got just what you wanted. 'Chance favours the trained observer', is an expression commonly used in such circumstances.

There have been many instances of this. The whole range of coal tar dyes was invented following the accidental breakage of a thermometer into a beaker containing tar extracts and a solvent, releasing mercury which acted as a catalyst. Although one might argue that they would have probably been invented anyway sometime later, it is unlikely that anyone would have actively looked for a whole range of colours in such a messy and intractable substance as coal tar. Again, Kerkule dreamed of the six-carbon ring in the form of a snake eating its tail, and found it explained otherwise unsatisfied loose ends in his compounds. The coming generation of self-healing, recyclable plastics is a more recent example of an accidental discovery.

Experiments, both real and imagined, should have some kind of theory behind them, and a good theory has predictive value. If the theory is right, then there must be certain things that will follow as a consequence. This means that a thought experiment can often be tested by means of its predictions, without anybody having to actually do the experiment. The theory of the Big Bang was a natural consequence of observations of an expanding universe, for if something is expanding one can ask the questions 'From where?' 'Where to?' and 'When?' We cannot make a Big Bang, however; it is on a bit too large a scale for one thing, and we don't have the ingredients to hand for another, but that need not stop us asking questions.

Anybody can join in this exercise and do Gedanken Experiments.

Quite often people are put off by the thought that they might be wrong and perhaps be exposed to ridicule. There are two answers to this. First of all, it is just as often the people who are right who are ridiculed. Secondly, the person who has never had a wrong idea has probably never had one that has been right either.

As Cardinal Newman said, 'A man would do nothing, if he waited until he could do it so well that no one would find fault with what he has done.'

It does not matter a whit if you are wrong on some counts, the object of the exercise is to have fun with ideas. In the following chapters I hope to show how to do this, to take a simple proposition, and then to add or take away bit and pieces of it, to turn it inside out and follow up minor side turnings that may open out to wide avenues. Quite often the Devil is in the detail, and just when you think you have explored a subject thoroughly, with just a few odd corners to be tidied up, you find doors in them that lead off to other dimensions.

I have been told that every mathematical equation in a book halves the readership. As I don't want to end up with half a reader, I have only used the kind of maths that you can do in your head. There are two reasons for this; one is that the book is supposed to be about doing things in your head, and the other is that I am hopeless at maths myself, and if I can't do it in my head I usually can't do it at all. Do use a pocket calculator though if you are used to it. If you are good at maths you can take many of the experiments a lot further, some of them into completely new ground. Also, you could find all the errors I have made by trying to do complex sums in my head, which you will probably find quite rewarding. I did say there is nothing wrong in being wrong!

I have also used pounds and feet and inches most of the time, rather than scientific units, for a lot people are more familiar with these, particularly in America, although even there the gallon is 20% less than the Imperial gallon. The teaspoon as a measure is bigger on the other hand, if you do cooking. You can use a conversion table if you prefer metric. In science of course the international system of units (SI) is always used, and is to all intents and purposes the same as the metric system. In engineering there is sometimes a bit of a mixture used, as NASA found to their cost when a Mars lander was programmed with a mixture of the two. It is very important in the real world to decide which system you are going to use and stick to it, with translation tables if necessary to communicate with others who may be used to a different system. I have tended to mix things up a bit in this book, but then, that is the real world!

In inventing, quite often it is necessary to go back to the beginning and start all over again.

To give an example of what I mean, I might be engaged, as an example, in combining hydrogen with oxygen in a fuel cell to produce electricity, with water as a by-product. Hydrogen yields much more energy per gram than any other element if combined with oxygen, mainly because it is so light, so it seems a good choice at first. However, I then find that it is extremely bulky and difficult or expensive to store in any quantity on a vehicle. Worse, the weight of the high-pressure cylinders I need to contain it is far greater than the weight of the hydrogen in them. We are halfway back to the problem with the lead-acid battery, where forty pounds of lead act as a carrier for a couple of ounces of hydrogen, which does all the work, so my weight advantage is eroding fast. The suppliers have a distribution problem; it is very expensive to liquefy, so they have to lay long distance insulated gas pipelines. It also leaks through the smallest hole rather well. The net result is that it is not worth the cost, so high pressure pipelines are used, which increases the leak problem. I could of course absorb it into palladium, but that is a very expensive and heavy metal, and if I put too much in, it is liable to explode.

Supposing I then have the idea that some elements have a greater affinity for oxygen than hydrogen, but don't give more energy per gram because the atoms are heavier. (Actually not my theory; it was all worked out in the C19th.) I can check this by putting a small bit of lithium, sodium, potassium or calcium in water. They all liberate hydrogen, sodium being the most vigorous and calcium the least. Thus the experiment confirms the possibility of making hydrogen from water on demand, rather than carrying it about in heavy cylinders. Conveniently, at the end of the reaction my fuel-cell produces water.

There are problems with handling bulk quantities of these metals however, apart from being a hazard in a crash, so something a bit less reactive might be a better compromise. Magnesium is stable, but a bit of a fire risk, however aluminium is in good supply. We won't run out, either, it being the third most plentiful element in the earth's crust after oxygen and silicon. One could use scrap metal; impurities won't matter and it is relatively cheap and safe to handle. It does not react with water though. The problem is that it forms a very thin but tough oxide film on any exposed surface, which insulates it from further reaction. The film binds on very tightly and needs to be disrupted.

Mercury will do this, but is expensive and poisonous. Caustic soda will do the trick though, and is cheap and relatively safe if one takes sensible precautions to avoid skin contact. As I want to generate my hydrogen more or less on demand I have to design some kind of throttle, so just chucking a load of milk-bottle tops into a tank of caustic soda solution is not a good idea, as I can't switch off the reaction. Thus it might be a good idea to introduce it a bit at a time. A wire feed (or ribbon, because it winds more densely and has a bigger surface area), might work, with a feedback control loop, which winds more in if the pressure in a small ready use tank of hydrogen drops. If I want to speed up the reaction a bit I could try passing a low voltage direct current down the wire as well, and reversing the current might provide a secondary throttle. I might even be able to reduce or do away with the caustic soda, which is corrosive and prone to foaming.

Here we have a whole string of Gedanken experiments, and a whole lot of further things to think about. For example, what do we do with the aluminium hydroxide mixed with caustic soda? We will produce it in great quantity if we design a transport system to run a few hundred million cars and lorries around using this idea. Luckily, this is exactly what the aluminium smelters want; they actually purify their bauxite by reacting it with caustic soda. So we can arrange the transport bringing the aluminium wire to the garages, to return full instead of empty, at minimal cost, to where they got it from. This would be a distribution point or perhaps a smelter running off excess generating capacity, from a wind farm or a nuclear power station at night. The energy cost is about 12-kilowatt hours per pound of metal, so we have a very good way of storing the energy in the electricity. To put it motorcar terms, a gallon of solid aluminium can yield much more energy than a gallon of petrol, and only weighs about three and a third times as much.

We could go on almost indefinitely, as ideas ramify, and there are a lot of cost benefit analyses against alternatives to be carried out before we decide this is the route we want to take. For example, ethanol cells are pretty good too, and here we have a fuel that can be tipped into an ordinary car during a transition period while we are gearing up to go electric. Also, what kind of cost reduction can we expect if we increase aluminium production, or might there actually be a price rise if we have to use less desirable ores? Is there another chemical route to using aluminium, for there is considerable loss as heat in reacting it with water; perhaps a direct conversion to a halide via an organic reactant might give more electricity? This would be an aluminium battery, which might be rechargeable. I read somewhere that someone had designed one in the thirties, but it used a lot of platinum! We might ask to what extent we could use a vehicle fuel cell to power rural homes. Long cables with low demand are lossmakers for electricity supply companies, the loss being hidden in the short run high demand end of the business, so they are actually paying more.

The process is now becoming one of invention, where there is a good commercial incentive to look at detailed technicalities rather than broad ideas.

There is nothing wrong with this; in fact it is just as well that innumerable people over a long period of time have engaged in this type of mental activity, for that is the basis for our civilisation and the basis for the wealth that makes it possible. Civilisations only spring up where there is surplus time or wealth, which is really the provision of goods and services divided by the time taken to create them, over and above that needed for pure subsistence. All wealth is the product of invention. 'Natural' wealth, such as oil, is not worth much if you do not have a lamp or an engine to put it in. A field of naturally grown grain, itself a result of thousands of years of selective breeding, is pretty to look at, but if you have ever tried eating it straight off the ear, you will realise it is strictly for the birds, and appreciate the invention of the mill and baking oven. It is perhaps instructive to reflect that half a dozen types of birdseed power this, the greatest civilisation there has ever been!

We tend to take our common heritage for granted, and to belittle the efforts of past civilisations. After all, a year's work for a whole tribe at land clearance for agriculture, five thousand years ago, could be done by one man with a big tractor in the space of a few hours nowadays. However, if you work out the compound interest on the value of their work, at only a couple of per cent, you will get a different perspective. Their success cannot be disputed. Somebody had to start agriculture, others had to continue it, to experiment, to select new crop varieties, to trade and invest the results of that trade. What they produced we have now; a massive inheritance that we have not worked for. Most of our food crops and many of our medicines are in fact Neanderthal in origin.

I will next look at something of less use, but which is more fundamental. There is always a kind of tug of war between pure and applied thought. The purists tend to take the high ground, pure research being presented as more 'fundamental', leading to greater knowledge and a broader understanding of the universe. Applied research on the other hand is presented as being money grubbing, leading to pollution and annoying things such as other people's cars and mobile phones. Unfortunately there seems to be a lot more funding for the application of technology than for the more abstruse lines of enquiry, unless there is great national prestige involved. Even national prestige is a waning force nowadays; the cost of nuclear research and space exploration is becoming so high that joint international initiatives are the order of the day, and even then there are cutbacks.

We will start with a very simple question, that small children will sometimes ask;

"If I dug a hole right through the middle of the Earth, where would I come out?"

If you are in England, you will probably say, "the Antipodes", which are a small group of islands a bit to the East of New Zealand. Actually, you would probably come up at the bottom of the Pacific, for the footprint of the Antipodes is rather smaller than that of England. If you dug a hole from most places on earth you would most likely end up in some ocean or other, for the simple reason that the surface of the planet is mostly water, and most of the land is situated in the Northern Hemisphere. This is no great problem however, for you could start at the far end; the rock dug up would make a fair sized island; after all, you would have a column eight thousand miles long and perhaps six feet across, mostly iron, to dispose of.

If you want to know accurately where you would come up, it is simple to find out where your antipodean point is. Merely look up your latitude and longitude in a decent atlas, or a globe if you have one. Latitude is measured in degrees from the equator, which is zero, rising to 90 North at the North Pole and 90 South at the South Pole. Merely swap North for South and you have your antipodean latitude; e.g., if you are currently at 45 degrees North, then your new location will be 45 degrees South. This gives you a line, but you also want a point. So take your longitude; this is measured from Greenwich, which is nought, rising to 180 degrees West if you go West or, 180 degrees East if you go East. There is no fundamental reason for zero being at Greenwich, unlike latitude, which is aligned with the spin axis of the earth. It has been at several other places in the past; it is merely a matter of convenience to have it somewhere that everybody can agree on. It was in Paris before that, and the Ancient Egyptians had it going through their Great Pyramid. As Britain became a naval power, and made a lot of maps it ended up here. If you add (or subtract) 180 degrees from your longitude, where this line crosses your new latitude, is where you will come out, and it is probably underwater.

So, problem solved. Quite simple really: there seems to be no more to do.

It is a great pity to waste that beautiful hole though! So what could we do with it? One good idea might be to jump down it. Best jump in headfirst, like a rabbit, for then you will come out headfirst and be able to step onto the side to take possession of your new island, or say hello to the natives if there is land there already.

How long will it take though? Will we need a book to read, and plenty of sandwiches? After all, it takes ages to fly to the other side of the world, a day or two at least.

Actually it would take about forty-seven minutes. I say about, because that is the figure I read in an article. (Since writing this, the time has moved to 42 minutes, then 38 minutes.) When I tried to work it out myself I made it longer, and so did a friend. However, I did not make allowance for the middle bit of the earth being more dense than iron; it is nearer fourteen grams per cubic centimetre than seven and a half, so would attract more strongly. Also I cheated by saying it is four thousand miles to the centre of the earth, whereas it is actually a few dozen miles less. I also assumed that gravity would start to reduce in strength as soon as I went through the surface, whereas it would rise slightly while going through the light rock of the mantle, only declining when I got into the iron of the core.

During the whole of the time of transit, you would be in free fall; i.e., a zero gravity environment, like being in a satellite. In fact, a line straight through the earth can be considered as a species of orbit. Orbits are nearly always elliptical to a degree, be it ever so slight. The planets, the stars in the galaxy, and our home made satellites generally tend to follow nearly circular orbits, but comets show much greater ellipticity. There are two extreme cases of ellipticity. One is the circle, where both axes are equal. The other is a straight line, where one axis is zero. You can only have this kind of orbit if you have a hole in the gravity field, like a doughnut, however; otherwise you will crash into the body attracting you!

On the way down we can do some experiments. Supposing we put lights in our tunnel, evenly spaced out, as in the channel tunnel. As we start to fall, we will find that they go by faster and faster. We will experience a gravitational field at the surface of the earth that will give us an acceleration of 32 feet per second, per second. This means that we will be moving at a speed of thirty-two feet per second after the first second, and sixty-four feet per second after the second second, and so on.

As we start with a speed of nought feet per second, and end up doing thirty-two feet per second, we will only fall sixteen feet in the first second. We will still have a long way to go, but things speed up quite rapidly. After the first second, we will start with a speed of thirty-two feet per second, and end up going at sixty-four feet per second. To find out how far we will have fallen, we add the two velocities, and divide by two, as this is the average velocity over the second. We will have travelled forty-eight feet in that second, plus of course the sixteen feet in the first second, making a total of sixty-four feet. If you are more used to thinking in miles per hour, a rough conversion is to divide the speed in feet per second by two thirds, to give miles per hour. For example, a speed of sixty miles per hour is 88 fps exactly; call it ninety and divide by two thirds and you have 60 mph. You can go on adding your speed up second by second, and will soon find that the acceleration is a lot better than the family car! Eighty-five miles per hour in four seconds sounds quite good!

I sometimes think it would be a good idea to calibrate speedometers in cars in 'feet above ground' as well as miles per hour. Most drivers tend to think of sixty miles per hour as being twice as fast as thirty miles per hour, which of course it is. However, they tend to assume, if they assume anything at all, that a car at sixty has twice the energy as one at thirty, which it most certainly does not! This is why stopping distances get inordinately long at high speed, and also a high speed seems to 'flatten the hills'; the loss of speed occasioned by a thirty foot rise when travelling at seventy miles per hour (the legal limit on motorways and dual carriageways in the UK) is far less than when doing a mere thirty. To double the speed of a moving object, you have to give it not twice, but four times as much energy. If you really put your foot down, and travel at one hundred and twenty miles per hour (legally) on the German autobahn, or (illegally) on the motorway, you will be carrying not four times as much energy as when you were doing thirty, but four times four, or sixteen times as much. If you put the brakes on you will take sixteen times as long to stop. In 'feet above ground' terms this is about five hundred feet, as opposed to twenty miles per hour, which is about sixteen feet, and if you hit something solid you will do about the same amount of damage as falling that far. I have travelled six hundred miles per hour on a plane, at nearly two hundred miles per hour on a train, one hundred and forty miles per hour in a car, and over a hundred on a motorbike, and it is not all that exciting. It does not look all that different from more normal speeds, so it is easy to have a false sense of security.

You won't have a false sense of security now though. Five hundred feet up without a parachute! People who read this book will probably have a slightly longer life expectancy, and be late for appointments more often, through dawdling.

To continue our free fall, after a few minutes of watching the lights go by to measure distance and working out speeds, you will find that things don't quite add up. Your rate of acceleration will be decreasing. Obviously once you have passed the centre of the earth, you will begin to decelerate, or slow down. It is perhaps reasonable to suppose that this won't be a sudden flip, but a decrease in gravity in one direction, to be followed by an increase in gravity, now retarding you, in the other direction. Suppose you decide to measure the force of gravity at the halfway point to the centre of the earth, two thousand miles down your hole. You might assume that as the sphere beneath your feet is now only half the diameter, it will only attract by one eighth as much, there being only one eighth as much mass in a sphere of that size. You would be right if you measured the field caused by such a sphere from your previous position at the surface of the earth, but of course you have moved to be on the surface of the new sphere as defined by the half radius. The problem is further compounded by the fact that the centre of the earth has a much greater density than the rock of the mantle; about fourteen as opposed to about two and a half for a lump of granite, or seven for iron. My Kay and Laby table of physical and chemical constants (only the 14th edition I am afraid!) gives the acceleration at this depth as nearly the same as for the surface, but I am not sure that is right. One thing I have learned in a long life is that constants tend to change over time!

You might wonder about all that mass above you. Will it not pull you upwards? Or at least have some effect? This is best examined once at the centre of the earth, for then everything really is above you, but it is above you in all directions, rather like being at the North Pole and finding every direction is South, so it will of course cancel out and there will be no net effect. Surprisingly, if you dig out a big cave at the centre of the earth, there will not be any net gravitational effect at any point in it. You could remove the entire sphere at two thousand miles radius, and float about freely at any point within it. You might think you ought to stick to the nearest bit, because it is, well, nearest, but you would be wrong. The reason is that there is only a little bit nearer to you compared to all the rest that is further away, and the net effect is a zero field. You can work it out with a vector diagram if you wish. To do this, draw a circle; forget about a sphere, a circle will do as well, and then draw lines, where the distance of a point from you, represented as a dot, is in the inverse square of the distance from the inner surface of your circle. You will find that you are indeed pulled quite strongly by the bit nearest to you, but the much larger number of points at a greater distance, pulling you less strongly, will cancel this out, by pulling you in the opposite direction.

Interestingly, we can extend this idea of a shell of material, so long as it is of equal thickness in all directions, having no net gravitational field inside it, to the universe at large. You experience no net gravitational effect from the entire universe around you, only that bit between you and the centre of the earth. This is not quite true, for the universe is not homogenous; there are galaxies that are not evenly spaced out and we experience a variety of very slight pulls in different directions, proportional to their mass and distance. As gravity diminishes as to the inverse square of the distance, the effects are actually exceptionally tiny, unless considered over billions of years. The same principle operates inside a black hole also. The net field at the centre must also be zero! This is one argument that matter is not compressed to a point at the centre of one. Probably it can only be compressed to a Planck distance, which is extremely small, but not zero. Try to compress it a bit more and it would probably start to expand into other dimensions, rather as a fistful of jelly would squeeze out between your fingers if you tried to compress it into a smaller volume.

Within our galaxy the effect is more noticeable. We are used to the idea that the planets near the sun go round rather quickly compared to us, while those further out take a lot more time to complete an orbit. The reason is of course that those nearest the centre don't have so far to go, while having a strong gravity field to hurry them on, and those far away have a much further distance and a weaker field. They not only have further to go, but they are going more slowly in real terms also. Mercury for example takes roughly three months to orbit the sun, while Pluto takes two hundred and forty eight years. The mass of the sun is such that we can almost neglect the mass of the planets in our calculation. This is not so for the galaxy as a whole. Although there is a concentration of mass at the centre, probably in the form of a black hole of many millions of star masses, most of the mass is in the rest of the star field. This means that it rotates at a much more uniform velocity, with the speed of stars at the edges being not that much slower than those nearer the centre. We would get the same effect if we drilled a number of circular tunnels around the world at different depths to our hole through the centre of the earth, and set masses into orbits around them. They would go slower than the speed necessary to achieve a low orbit from the surface.

Actually, there is probably a lot more mass outside the visible extent of the galaxy, and this also will tend to make it rotate at the same rate, as though it were a flat solid sheet rather than a planetary system. Most of the mass is 'dark matter'. Dark means invisible, so nobody has actually seen any yet.

I am not quite sure about the idea of a net zero field. Is this the same as no field at all? Or is there some kind of gravitational tension in space? To illustrate what I mean, I used to run quite big midsomerfest barbecues for foreign language students, mostly Scandinavian, as I had a four acre garden. I had an old piece of hawser, salvaged off the beach by a friend. It must have been used to moor some large ship, as it was about ten inches (ropes are measured by the circumference) and several hundred feet long. It was pretty old as it was made of hemp; most ropes are plastic now.

Anyway, we had a tug of war, girls against boys, over a hundred of each. To do this, the rope was draped over two garden benches a few feet apart, with a handkerchief tied midway between. Everybody took up tension, the handkerchief was adjusted so as to be midway, and the signal was given to pull. The boys dug their heels in and grunted, the girls leaned back gracefully and tittered, and the handkerchief moved not at all. Now there was a considerable difference in the tension of the rope before everybody started to pull and afterwards, but it had no effect on the handkerchief. There was no net observable effect.

It would be interesting to see if there was any difference in the speed at which a clock ran in a gravity free, compared to a net zero gravity, environment. Perhaps someone reading this will have access to a LaGrange point, a kind of backwater caused by the interaction of the gravitational fields of the sun, earth and moon. We could put an atomic clock there and compare it to one on earth, and one sent a long way out of the solar system. Gravity slows clocks down, so we might find that a state of gravitational tension is cumulative for clocks, even though two fields cancel so far as weight is concerned.

Oh! Why did the boys not pull the girls over? Well, the girls had led their end off into the bushes, and tied it to the base of a tree. Girls are like that.

Having found a very quick way to get to the Antipodes, we might decide to go somewhere more interesting. Like Nice, in the South of

France perhaps, which is not very far by air from Southampton, near where I live. Unfortunately I have to go the wrong way to Gatwick first, by slow train, changing trains several times so it takes hours and hours. I then pass over my house and am in Nice an hour or so later. Why they don't develop regional airports, or at least feeder lines from them I don't know.

How long would it take by digging a hole? Perhaps it will be much quicker, as it only took forty-seven minutes to get to the other side of the world? However, it may be slower, as our hole is only at a very shallow angle, and we will not accelerate nearly as fast, and the slowing down bit will be just as tardy as well.

Actually, it will take forty-seven minutes! You want to visit the people next door? Forty-seven minutes! The shop at the end of the road? You guessed it!

However, suppose you wanted to travel somewhere a hundred miles away? Forty-seven minutes seems quite good, and although difficult, such a friction free, vacuum tube railway might be in the bounds of future possibility, which one through the centre of the earth is not.

There is better yet, however. Supposing we only want to go a few miles and we can reduce the friction to an acceptable level over that distance. We don't have to dig a straight-line tunnel; we can dig straight down for a few hundred feet, put in a curve that would give tolerable 'g' forces to turn through ninety degrees to the level, and proceed in a straight line with a reverse turn at the end. Hey presto, gravity powered travel at over one hundred miles per hour with very rapid acceleration and all the braking energy saved up for the next trip. This is still not totally practical. I don't expect people would appreciate free fall for several hundred feet, even if it did teach them not to speed on the motorway. However, we could dilute gravity by, say, a one in five slope and still arrive pretty quickly.

We seem to have got back to speedometers marked in 'feet above ground' again! There is a reason; these essays have links and we will return to the theme in an article called 'What about the Railways?', where we will have stations with platforms stacked one above the other, to save acres of space, and gain energy and save braking loss, plus of course a lot of other good things that could easily be put into a rail system.

So far we have got quite a lot of mileage out of our hole through the centre of the earth. There are lots more interesting things which we can extract. For instance, we have not mentioned coreolis force. This is why the water spirals down the plughole, and depressions spiral anticlockwise North of the equator, and the other way in the Southern Hemisphere.

For simplicity, supposing we dig our hole at the equator. If we look at the earth from above the North Pole, we will see that it is spinning. As we enter our hole we will have a velocity of one thousand miles an hour in a West to East direction, because it is twenty-four thousand miles around the equator, and we do it in twenty-four hours This is why the sun appears to rise in the East; we are going East at whatever velocity our latitude is going, unless we are at one of the poles, in which case we are merely rotating. However, when we come out of the other end of our hole, we will be going in the opposite direction. Well, we will still be going East, but we are now on the other side of the world, so if you look at it from above the pole you will see that it is in fact the other way to the direction we were going before. So in our forty-seven minutes we have slowed down from one thousand miles per hour at the surface, to nought at the centre, and back up to one thousand miles per hour in the other direction when we re-emerge. This of course is in addition to our speed of fall. This means that we have experienced a lateral acceleration; the sides of the hole would have pressed us sideways as we made our transit.

This acceleration due to rotation is often called a centrifugal force; if you whirl an object round your head on a string, you can feel it in the tension of the string. Strictly speaking, the force is centripetal (towards the centre); however, bowing to common usage, I will call it a centrifugal force. It means that the world bulges out a bit at the equator, about thirty miles I believe, compared to a perfect sphere. It also means that if you mine gold in Alaska, and have it weighed again further south, you will have lost a bit of weight, against a spring balance that is; a beam balance uses weights which also lose or gain weight in proportion, so that is not affected. A beam balance is really a means of comparing mass, which is inherent in the nature of the atoms, rather than weight, which is merely the effect produced by the local gravitational field, or acceleration. At the equator you will weigh a few hundredths less than at the pole on your spring balance bath scales.

You might want to work out how much this acceleration is; it is quite easy to do so. To do this convert your speed into feet per second; as you have experienced a change of one thousand miles per hour in one direction, to the same in the other, this is roughly three thousand feet per second. (Remember the rule of thumb conversion of miles per hour to feet per second.) Then find out how many seconds you took to attain it, and divide. Your sixty seconds in a minute conveniently goes into your three thousand feet fifty times; cheat a bit and call your forty seven minutes fifty, and you have an acceleration of one foot per second. This is only a rough approximation due to the short cuts we took in calculation so as to be able to do it in our head. If you work it out more accurately it is still only 1.32 ft sec. This is not very exciting, so you will probably hardly notice it. If you weigh a hundred pounds it will be equivalent to a force of about 4 pounds pressing you sideways.

The above might be classed as a completely useless exercise; personally I prefer the term inutile, not to be confused with futile. This is not the point however; it is interesting, to me at least, to get a new perspective, and maybe the information will come in handy someplace else.

I will not continue with this example now. The idea is to show how ideas can be extended, by almost a sort of random walk, so that you end up in a different world from the one you started in. All the results can go into your Gedanken drawer, where they might come in useful later. We will look at a hole in the earth later as a possible route to getting deep earthquake sensors. It is quite likely that you may turn up ideas that are commercially viable, or you may be an inventor already; everybody probably has at least one invention. It has been said that there are three ways of making money; wine, women and inventions, and that of the three, inventions are by far the most lucrative. All the wealth in the world is the product of invention. Unfortunately, it is seldom the inventor that ends up with the money! Other people know this, and the road to invention is beset by highwaymen.

Most inventors invent for the sheer joy of the thing; problem solving is innate in human nature it seems. There is considerable satisfaction in finding a solution, and this is particularly so when you know you are the first person to do so. The fact that there is no immediate or prospective use for your new idea is not the point. The laser was described as a solution waiting for a problem when first invented, but it seems to have caught on a bit now. I met Professor Gabor shortly after he cracked the problem, and I think he was more interested in solving the problem than its uses. I had a previous interest in holography from my photographic days. A sheet of printing paper could be had very cheaply after the war, as there was a lot of not quite time expired Government surplus about. I realised that a sheet of paper, exposed without a lens, actually had as much, if not more, information from the light falling on it as one exposed with a lens, but jumbled up. All one had to do was unscramble it. I tried exposing a sheet with a box of a gross of milk straws against it as a collimator, to get a kind of bad 'insect eye' impression, and briefly tried to unscramble it with a lens after the event. A bad idea! Eventually I decided that the problem was impossible without a source of coherent light. Of course, someone else actually did make a poor holograph using ordinary light later, so I not only failed, but proposed the wrong reason for failure also!

After the first computer was built the market was assessed as 'about three', but they have sold a few more since. I will include one or two problems later on which are of no use whatsoever, other than that they might be fun to solve, and present a learning curve on how to solve problems that can be transferred to other ones that might be of more use.

I will also give one or two examples of 'impossible problems'. There is great merit in trying to solve impossible problems, so long as one does not get obsessed. There is no possibility of failure if attempting an impossible problem. You will be certain to get the expected result! What you may do is to get it by a route nobody else has thought of; also you may get an insight into other unsolved problems which are actually soluble, but need a different approach. Again, all problems are insoluble until solved, and who knows, you might get lucky and crack some puzzle that has been around for a long time. The solution of Fermat's last theorem is one such example. Also, it is certain that he would not have done it that way. He may have been wrong, of course and not have been able to do it at all. However there is probably a much more elegant and shorter solution potentially findable. Just think, you might be the second person to solve Fermat's last. I would not seek to encourage you though, mainly because I had difficulty understanding the question when I first met it, let alone the answer. I don't think it could be done in the head, although Fermat probably did, or at least saw the main way to go without putting pen to paper. Sometimes a long solution to a problem can trigger a succession of simpler ways. There was a problem concerning the least number of changes to work through an exercise in bell ringing; I have forgotten the name, probably a triple Bob Martin or some such, bell ringers use that sort of name. It was tried on a computer in the early sixties, but proved too hard; then someone solved it in many pages, to be superseded by a much shorter version, to be trounced in turn by someone who jotted it down on the back of a paper bag. Quite often without the first solution the others might not have occurred.

I hope to present a mixed bag of topics and problems, with something for everyone. If you are the technical sort, whose primary interest is in the nuts and bolts of a real life problem, do take a quick look at the more theoretical problems. Conversely, if you are interested in ideas rather than mechanisms, do look at the mechanisms too, for each field can fertilise the other. Most chapters are in the form of a number of short essays, with some explanation and development. Many of them are linked, or have common elements. Most have not been explored in exhaustive detail, so there is plenty of room for them to act as a starting point if you wish to continue. However, once you have got the idea, you can branch out on your own; there are all kinds of discoveries out there waiting to be made. The doubling time for scientific knowledge has been given as between five and seven years and shortening. After all, we have only been going with scientific method for a few hundred years, most of that done in the last century. The amount that has been discovered is an infinitesimal amount of that which is waiting out there. For all we know the universe of ideas may be infinitely complex even if there are physical limits.

A background in Science or Philosophy helps, but you can follow up by reading; one book leads to another and you will soon have an area of expertise. I have mixed feelings about reading all there is to know about a subject before working on it. Maybe it is just the way I work, but I find too much knowledge can be inhibiting. It is only a small step from knowing all that is known about a subject, to believing that you know all there is to know; a fatal mistake!

For this reason I have not put a bibliography at the end. With the best will in the world, it would channel your enquiries in a certain direction, whereas if you select your own reading you will follow your own inclination, and hop around, which may be more suited to the way you think.

HOW TO PROGRESS AN INVENTION

This section is mostly about practical matters, with some inventions you might like to develop thrown in by way of example, although you will probably have plenty of your own.

In this chapter I wish to show how an invention can be developed. The starting point for an invention is often an unfulfilled need. Someone is engaged in or observes an occupation or activity and sees that it is awkward, or involves too many stages, and could be improved. There is usually an initial idea, which is tested either by practical application or by thinking through how it would work; this is often followed by a modification which is then tested, and a process of 'honing' takes place. The object of this may be to improve the design for ease of use or manufacture, or to reduce cost. Eventually a final form is arrived at for production. This will usually be modified again later, although much more slowly.

The reasons for this tailing off are several; there will probably be less leeway for later improvement, much of the improvement having been achieved. The cost may escalate against decreasing gains, while a large amount of capital may be tied up, leading to conservatism regarding change, even if desirable.

I will give an example, using a holistic view of an industry, which splits into several related inventions. Aeroplane speeds have largely remained static since the invention of the jet engine. The reason is that the cost of increased speed is not linear; to double the speed requires much more than a doubling of fuel consumption, thus the payload decreases as the cost of operation rises. Also, while a doubling of speed from 250 mph to 500 mph will save 2 hours on a thousand-mile journey, doubling the speed again will only save one hour; half the saving for about four times the cost. Working in favour of an increase in speed is the fact that the capital involved will be used more profitably. A million pounds worth of plane that can cover twice the fare paying distance halves its capital cost per mile, for example. However, it will still take just as long to load and unload the baggage and passengers, while servicing may be more expensive. In this case the market will reduce also; there will be far fewer passengers willing to pay the increased cost.

More importantly, there is likely to be tunnel vision in the industry. It has been said that the proper study of man is man. It seems to be that the proper study of the aviation industry is perceived to be aviation. However, what the paying public wants to do is to travel, and this embraces the trip to the airport, probably a couple of hours, getting there early as like as not to allow for motorway hold-ups that may or may not occur. Then there is the wait for check in, commonly two hours. There will be a long walk to the plane, or maybe a bus, and a slow loading time while the person in front puts things in the overhead rack, only to take the outside seat and get in the way. There are then two miles of taxiing, and a wait for the control tower. Having spent four hours in waiting and travelled eighty miles in the wrong direction, the passenger is finally off. An hour later he has a wait of an hour to retrieve baggage and clear customs, with perhaps another hour's travel to his final destination, seven hours and four hundred and twenty miles from home. The average speed is about sixty miles per hour, although the plane did five hundred.

Obviously there is little point in designing a plane that does a thousand miles per hour for this kind of journey. The logistics get a bit better for flights of several thousand miles, but not a lot; the fundamental errors of perception are still there. The way forward is to think much more holistically and embrace the whole travel experience, not just the minor amount of time spent in the air.

To do this, regional airports should be developed, so that travel time to the airport is reduced. Every airport should have a railway station within it, and a motorway connection, with bus station, plus cheap or free parking. Check in times should be reduced to twenty minutes or less, boarding facilitated by having more exits and lining passengers up by seat numbers so that the first in the queue are the furthest from the entry points. The embarkation point should be as close as possible to the take-off end of the runway to avoid interminable taxiing. Some airlines still bring round 'duty free' items for sale. There is nothing more brain-dead than carrying unnecessary payload. People could still pay for items on-board, the order then being radioed ahead for pick up on arrival, which is now being done more often.

There would be a good case for charging on the basis of total weight, i.e. passenger plus luggage. There is little merit in charging a nine stone weakling a huge amount for excess baggage, while a twenty stone person has twice the weight transported for less. Baggage check-in could be greatly facilitated by pushing sales of 'airline luggage', produced by the plane-maker, airline, or subsidiary. This would be lightweight, guaranteed to stow neatly as baggage in the hold, or fit the overhead for small items. A built in scale in the handle would show on a dial the exact weight of the piece when full. Fast track check in could be provided for such baggage, which would greatly enhance sales. It would also be designed not to tumble on the carousel. With slimming margins airlines should not miss a trick. 'The package is part of the package!' There are many points where a small profit may be gained with advantage both to the operator and the customer.

Baggage loading and unloading should be unitary; i.e. a single container holding all the luggage should be loaded in a matter of minutes, rather as an airborne lifeboat used to be loaded in the bomb bay of a Lancaster during the second world war. The container itself should be designed to discharge baggage, via rollers, in a steady stream onto the carousel conveyor. No more throwing things about individually by hand! The above package of ideas would reduce transit time to less than half for most journeys and cost much less than faster aircraft, which would not do the job anyway.

The classic example of tunnel vision was Concorde. From the outset it was known that it would never pay; also it was not a particularly safe aircraft due to its very high take off speed. Even to the last the tunnel vision persisted. The plane actually flew quite well. Had I been asked, I would not have taken them out of service, but have retrofitted them with modern, lighter, more fuel efficient engines of less than a quarter of the power, and flown it as a subsonic airliner with a greater range, payload, and lower take-off speed. Rebranding would have been necessary after the crash. 'All new,

Concorde Two!' would have done the trick. A whole new market would have been opened up; as a special treat the passengers could have been offered their cutlery set for a tenner, thus getting another nine pounds profit, and reducing the amount nicked!

WHY IT CAN'T BE DONE

Quite often, when proposing something new, you will find friends immediately start to raise objections. These objections will be of the most ignorant kind, showing their lack of understanding of the subject matter, an unwillingness to change, and an assumption that you are the enemy, and need to be stopped. There will also be an element of assumed superiority, as though they know far better than you. Do not take this personally; it is a kind of Pavlovian reflex, something unconsciously produced by the new and the strange, a sort of built in defence mechanism of the human psyche. Life was difficult enough in ancient times, what with famine, sabre toothed tigers and the like, so the safe old way of doing things was better than any newfangled notions.

You get the same thing in future studies; you are regarded as some kind of crank or unlicenced prophet, best relegated to Speakers Corner in Hyde Park. Prophets in ancient times were often told to become soothsayers in the marketplace; a lowly occupation. Mohammed was afraid he would end up there, and one of the Biblical Prophets (I forget which one, probably more than one even if not recorded!) was told to go there. It used to be said that a prophet has no honour in his own country; unfortunately, we all live in the global village now. The prophet however has drawn a shorter straw than the inventor; he quite often does not know what he is talking about! There is no requirement to understand, or even believe; all he has to do is to say the words only. At least the inventor usually has some facts and figures to back his idea up, or a model.

I will give an example of this sort of thing, by way of a letter.

DEAR MR DUNLOP,

Thank you for your copy of your patent at Kings' for your new invention of the inflatable rubber tyre. We, at our new Government department for Forward Link to Opportunities for Production. (FLOP for short,) will pass your invention on to our sister department, Finance And Invention Liaison (FAIL), with our assessment.

With regard to the invention, the principle characteristic of India-rubber is that it bounces. This would be undesirable in a conveyance or a bicycle, making for an uncomfortable ride. With the state of our present roads such bouncing would be more or less continuous, although fluctuating as the vehicle traversed the occasional smooth patch, then violent as bumps tended to reinforce it. On a corner there would be the probability of loss of traction and skidding.

Also, with the present state of our roads, which are largely flint or rough stone, there would be the constant problem of punctures, which would be a daily occurrence. This would occasion a greater loss of time in repair than would be saved by the bicycle in the first place. For four wheeled vehicles the problem would be twice as great, and any increase in speed would proportionately increase the risk. At present our stage coaches average twelve miles per hour, with eighteen being commonly exceeded on downgrades. The failure of a rubber tyre at such speeds would be catastrophic. I think it unlikely that many people will trust their lives to riding on four party balloons!

With regard to the material, India-rubber, despite its name, comes from South America, it being the sap of a tropical forest tree, tapped in rather the same way as maple syrup. These trees are widely separated, perhaps only a few in a square mile, thus the work of collection is very labour intensive, involving many miles on foot. Although the local population will do this for virtually nothing, productivity will be very low, so it would be impossible for the invention to have widespread adoption due to the shortage of raw material, which, due to transport and processing costs, will never be cheap. The possibility of growing rubber trees in plantations, such as is practiced for cocoa, tea or coffee, is not viable, for the cost of setting up such plantations would be prohibitive, especially as one would have to take seed (if one can get it) from an unmodified wild stock, whereas coffee tea, and cocoa have been cultivated and refined for centuries, if not thousands, of years. Despite this, they are still very expensive and only used by the spoonful. Rubber trees would also take far longer to mature, occasioning a long lead-time where money is deployed on the ground with no return. Nobody is going to plant rubber where there are greater gains to be made more quickly by planting the alternative. You could try dandelions though.

However, although your current invention is impracticable, there is a need for a new type of tyre. Foam rubber is better, but the restriction of supply still pertains. Linen bound rope, with or without springing under the saddle or front wheel, would be a much better alternative.

I am sorry if this is a bit of a disappointment to you, but far better to quit now and engage in more productive ideas before money is wasted on such a project.

Thank you for sending us your material.

Yours sincerely,

I. Stoppit. Patent examiner.

FOR THE ENTREPRENEUR

Although entitled 'For the Inventor', equally if not more important is the Entrepreneur. This is a person who can take an invention and guide it through the stages of modelling to see if it can be made faster, cheaper, or better, by refining manufacturing techniques, testing the market, and advertising, exploring overseas potential and so on. Personally I think this is a greater skill than inventing, although I once met an entrepreneur who thought the original act of creation was the more important skill, remarking, not unreasonably, that one had to have something to be entrepreneurial about first.. Personally, I am an inventor only and would not have a clue how to be an entrepreneur.

Here are some inventions you may wish to progress, from simple quick ideas to more fully developed designs that merely need implementation.

First, some quick fixes:

FOR THE SLIMMER

Overweight is an increasing problem, and there are marketing opportunities. Most very overweight people do not know what they weigh. Why? Because the average bathroom scales only go up to twenty stone; make a bathroom scale that weighs up to thirty or more stone and you have an untapped market.

Slimming aids sell; one sideways idea is to increase weight to start with to give a boost to the slimmer. Design a belt, rather like a money belt, with fourteen pockets to take one pound slim lead weights, or half pound weights if that is a bit much. When the slimmer has lost a pound they take another out of the belt. A double reward! Carrying an extra stone around focuses the mind on extra weight and weight loss. When they have lost a stone they can start all over again. Don't go sailing in it though. You could design a quick release like a car safety belt to go with it.

SLOPING TOPS

At the turn of the last century, the school inspector in Gosport could not abide the sight of piles of paper and books on top of cupboards, looking untidy and gathering dust, so he had all school cupboards designed with 45 degree sloping tops. There is mileage in this idea! Hundreds of people every year fall off chairs, trying to put things on or take them off cupboard tops. You could design an easily fitted adjustable sloping top for cupboards and achieve good sales whilst saving a lot of broken legs.

How do you avoid falling off your chair putting it up there? Easy! You get someone else to do it!

A BETTER MOBILITY SCOOTER

Mobility scooters are becoming more common, partly because of the ageing population, and partly because fat people are lazy, or at least less mobile. However, if you look at the models on offer, there is nowhere to put a sack of potatoes, and there are no models where an ageing partner or maybe a carer can sit on the back. So by a little redesign you could incorporate a boot for shopping, and a seat for a carer.

If you are into town design, you could incorporate a scooter lane as well as a cycle lane, or maybe combine the two. Once scooter lanes are established you could raise the maximum speed from eight mph (four on the pavement) to say fourteen, and make that applicable to cycles in town also.

I once borrowed a scooter which had handlebars and brake levers just like a bike, except that the brake levers were the accelerators. I expect they have been responsible for lots of accidents. How somebody comes up with such a brain-dead invention I don't know! Personally I shall get a ride-on lawn mower; they go faster and people are more likely to get out of the way in the supermarket aisles.

A COUNTING PEN OR PENCIL

Quite often, particularly in exam marking or tally work, a pencil is used to make a tick on the side of the page. These are then added up at the end of the page. The pencil could do this for you. All that is needed is a shroud which fits over the end of the pen or pencil, to form a grip; at the top is a two or three wheel counter actuated by a clicker mechanism (a toothed cam wheel that clicks round by one tenth steps), driven by a slight downward movement of the shroud. At the end of the page or paper one merely has to read off the number of ticks; write it down, push down a centre button to reset zero, and start again. The device could be sold with or without the pen or pencil. A locking clip could prevent counting when you did not want to.

IMPROVEMENTS TO CENTRAL HEATING

Virtually everybody has central heating nowadays, yet there are deficiencies that can be addressed. Usually radiators are painted white; actually they convect as much as they radiate, but white is the worst colour for radiation, black being the best. One may not want black radiators, but there are three other surfaces that could be black for the usual double radiator. Also, the valve is down at floor level, probably behind the sofa. They should be on top where they can be got at. Single lever half or quarter turn are best, not screw threads that take an age to operate. In addition they should have ballofix valves, so that an individual radiator can be isolated and removed, both the pipe-work and the radiator being isolated.

While on the subject; central heating has largely replaced the old fashioned fireplace, which for all its shortfalls did give very good ventilation. The problem could be largely offset by a triangular pipe and silent fan in the corner of a room to circulate the air, giving a mix of warmer and cooler air from top to bottom. A filter and or ioniser extractor could be incorporated. More extensively, a heat exchanger could be incorporated and exchange with the outside air embraced. And another thing: how often have you checked a radiator to see if it is on? It should be simple, but it is buried behind the sofa, or it is across the room and you have just sat down. A small mobile that sits on top of it or sticks on the wall over it would give a visual indicator, as the hot air would make the mobile go round; you could check without getting up or fighting the furniture.

AN IMPROVED TAP

Did you know that most taps are upside down? Usually the washer is screwed down against the water pressure. A far better arrangement would be to have the washer screwed upwards so that the water pressure helps close the tap. This would lead to better washer seating, preventing the edge of the washer being forced up, which can lead to channelling of the seat.

A QUICK DETACH OIL CHANGE

I once designed a quick detach oil change for my motorbike; it would however do for any engine. The usual process is to buy the litre or so of oil in a can. One then undoes a plug and drains the old oil out into a tray or some suitable receptacle. One then re-inserts the plug and tips the new oil in. What to do with the old oil? Why, put it in the empty can of course!

It would be much simpler if the oil tank on the bike was in fact the can you bought it in; a simple bayonet type fixing is all that is required. All you have to do then it to take the cap off the can, swap cans and put the cap back on the old one. Of course, many engines are wet sump. I will leave you to design the wet sump variant. While you are at it, arrange a can exchange system, new for old, so that the old can be recycled. It would still be useful unmodified for oiling door hinges or chainsaws. Also, it should be tax free, to encourage recycling by distillation.

IMPROVED CARD SECURITY

Everybody has a card of some sort nowadays, but did you know that they can be read remotely at a few feet by anyone with the correct equipment? You could design a range of attractive card cases or wallets etc., encased in fine mesh aluminium or other metal chain mail, or even a metallised fabric. Baco foil would do for a quick fix if you want one now. Uptake should be quite good as people are becoming concerned about security, mainly because most people have been ripped off at some time. If you are one of the lucky ones, it won't last. The problem is aggravated by the fact that many financial institutions do not bother to pursue small amounts. Consequently, if a card or its number is stolen the thief will make a series of small withdrawals rather than a big one. You could sell several million worldwide; the de-luxe version could give an audio signal to let you know you were being scanned.

Mobile phones are often stolen. You could design a transparent stick-on for the face with a picture of a crack or star impact on it, nobody will want to steal your phone then!

A MESSAGE SERVICE

How often have you called on someone to find that they are out?

It would be convenient if you could leave a message. To do this a dictaphone would be convenient, or a keypad so the visitor can type in a message, or a simple note pad and pencil in a waterproof bag if there is no porch. Such an item or range of items would find a ready uptake; you should sell a million.

FASTER WEAVING

Compared to many industrial processes, weaving is rather glacial. If you walk by a high speed loom, and return in a few minutes, not much will have changed.

Some palms have a kind of cloth, a bit like sacking or Hessian, sometimes called bast, formed from old leaf bases, and native peoples use this as cloth. On closer inspection, you will see that the cross strands are not woven however, but they are laid one on top of the other and fused at the crossover points.

For plastic threads, this technique could be used to speed up weaving. The warp and the weft could be laid, unwoven, one on top of the other, and then fused by pressure or heat treatment. The process would be several hundred times faster, as tens of thousands of crossovers could be fused at once. In practice thread could be continuously spooled or extruded for the warp, while the weft could be laid across several hundred threads at a time, again being spooled or extruded. A press or heated roller could then seal them together. For threads containing carbon fibre, an electric current could be used; a kind of low temperature welding.

OF GREASEBANDS AND RAILINGS

Some people grease-band their fruit trees to stop insects climbing up. In my experience ants are the worst. They farm greenfly up high out of reach; if you see a stream of ants running up and down your fruit trees it is too late. I don't know where the greenfly go in winter, although I did find some over-wintering on the underground stems of Rudbeckia once, so if you have both, dig a bit up and see if they are there. I expect the ants put them there as well, and possibly on other plants.

The problem with grease bands is that with a fissured bark they can get underneath, so you could sell double sided grease-bands, like flypaper. Personally I use car grease; you can buy it at the garage in little tins of a pint or so. There is a washer with a half inch hole in the top, so you can fill a grease gun without getting it all over the place. I had a friend who tried to fill my oilcan with grease once; he had a degree in engineering too! I don't know what they teach them nowadays.

All you need is a rubber glove; push the washer down to get a handful, then wipe it up the trunk. Do a couple of feet and it will prevent rabbits nibbling the bark and killing the tree as well. I taught a friend the trick once. He had planted an orchard and had a big rabbit problem. A few cans of grease did the trick, until he had two feet of snow that is!

The same rubber glove trick works on railings too. If you have ever painted railings, round or square, you will know what I mean. It takes ages to paint all the way round. If you try to do them all from the same side to save time, you get paint in your hair trying to see the back, and always miss bits. Never again! Get a sponge and a rubber glove and all you have to do is to dip the sponge in the paint and rub it up the railing. A much better job at ten times the speed.

You could sell a 'Paint your Railings' kit, with instructions. Don't tell people they will get paint on their boots instead of in their hair though. You will need to develop the invention; a glove with sponge on the fingers, or, more simply a sponge with a rubber band to go round the back of the hand might serve.

BETTER CARPETS

Most carpets are too dark, resulting in a fair slice of the money you pay for lighting being absorbed into a dark sink. Maybe the idea is to hide the dirt, but a suitable pattern can hide dirt on a light carpet. If you are making or selling carpets, state the albedo, which is the amount of light reflected. Make a sales pitch of it; you could say how much electricity it would save a year, and how many years it would take to pay for itself. The same could be applied to wallpaper; in fact, where appropriate, all paints could come with an albedo rating. For the really keen interior designer a colour spectrum could be added as well.

A TOP BOX FOR BOATS

Most motorcycles and scooters have a top box on a rack aft of the pillion. Apart from making a backrest, it is a useful way of carrying shopping, waterproofs and so on. Some are detachable and can be used as a suitcase.

A similar box would be useful on a boat, particularly for carrying an outboard fuel tank, which is the last thing you want to stow in the bilges. It could bolt on the aft deck and have a drain hole for any spilt fuel. Other uses would be for ready to hand items you might need to reach; a heaving line, spare lifejacket, a bucket for fishing, and so on.

AN IMPROVED JIB CLIP

While on the subject of boats, the foresail or jib is usually clipped to the jib-stay by a series of clips, which have a pull out knob on a spring. A simpler method I devised was in the form of a lazy key ring that is not tightly closed, so it can be pushed onto the stay and rotated until the stay is captured; the ends of the wire are bent into loop for ease of handling. The advantage is that it is cheaper, quicker, and can be used one handed; quite important on the foredeck of a boat that is bouncing about, as it leaves one hand for the boat.

A SECTIONAL ANCHOR

Another improvement would be a sectional anchor. Sometimes

you may wish you had a bit more weight down, particularly if it is not good holding ground with a strong tide. This can be achieved by have extra flukes that can be added to a normal Admiralty pattern or other anchor, the flukes attached to the outside of the existing ones in echelon. You would not want to have them permanently rigged due to the extra weight and awkwardness, but it would be handy to beef up your anchor if required.

MOORING BOOMS

In crowded harbours, boats are usually pair moored. The problem is that they can rub together, damaging the gel coat or paintwork. Normal fenders ride up and end up on deck, while long ones with a rope each end can still cause problems, particularly if they get tar on them.

If you look at old pictures of battleships in the early part of the century, you will see diagonal lines on their sides; these are stowed boat booms. When in use they would be mounted so as to stand clear of the side, and boats could be moored to them without banging against the side of the ship. Similar booms can be made of aluminium scaffold pole, which floats. A simple end-piece through which a rope can be threaded, ideally a split ring with overlaps, so the rope can be entered without threading the end, is fitted. The booms can then be mounted fore and aft to hold the boats a foot or so apart by bracing the mooring lines. If one boat goes off they can be stowed on the deck of the other; if both go off they will float ready to be picked up again.

A TWO SPEED GEARBOX FOR BOATS

Once, on the Thames, we met a guy who had put a car engine into his boat, and left the gearbox in. His idea was that you needed a lower gear to push upriver, while you could use a higher gear going down, when the current was with you. We laughed our socks off, of course, for the boat/engine speed is relative to the water, not the river bank.

However, upon reflection, he had a point, not in rivers, but while sailing. Quite often, in light winds, it is the practice to 'motor sail'; this involves leaving the engine ticking over at low revs while also sailing. There is a hidden advantage that by pushing the boat through the wind, you extract more power from it; your sails have a bigger slice of air to go through. Thus there would be a case for a higher gear for idling the engine under light load at a higher speed, giving a greater input from the sails as well as the modest push from the engine. I have on occasion been bumping into a heavy sea, directly to windward, when a lower gear would have been useful too.

EMERGENCY RUDDER LINES

On occasion, one may lose steering on a boat due to breakage of tiller or rudder lines. This problem can be easily solved by the addition of a small item screwed to the top of the rudder. It consists of a small metal hook on the outboard edge of the rudder, pointing downward. If a rope with a couple of knots is dropped over the stern, it can be lodged in this hook, giving replacement emergency rudder lines until the problem can be fixed. Nobody need go over the side and steerage can be regained, possibly with two helmsmen to make it easier if you have plenty of crew. If not, one line will usually be in tension and you can keep a hand on the other to prevent it trailing free.

JUMPING CATS!

I once designed, on paper, a catamaran with a deck like a glider's wing. The other wing could form the sail. When tacking, altering the flaps could cause the catamaran to jump off the surface of the water, leaving control surfaces immersed. It could then glide a short distance directly to windward, thus gaining ground on rivals which did not. Currently big cats are out of the water most of the time, riding on immersed planing surfaces, they cannot so easily jump to windward though.

A BICEP CLIP

An invention for the home handyman would be a bicep-clip; this is just like a bicycle clip except that is has a rounded section formed in the clip which can take a cable. Power tools are in common use and the cable always get in the way; by clipping it to the bicep it will be away from the work piece, and follow arm movements. If the tool is dropped it will strip off, so you don't get it dangling round your ankles. A plastic dip will provide suitable insulation and it could be tight enough to stay attached to the cable in the right place if taken off temporarily.

APE HANGER BARS

Most people get a bad back at some time or another. It helps to hang off something, taking the weight on the arms while trying to get the back upright. A simple bracket with a fold out bar could be produced; bolted to a wall high up, it could be swung back when out of use, or used by the parrot. Door frames are easier to fit to, but a bit low. It should be height adjustable for all sizes of backs.

AN IMPROVED FLOORBOARD NAIL

Another idea would be an improved floorboard nail. The current variety is all right, but one has to tap the next floorboard up against the one just nailed, with the occasional gap left by accident. The new nail would be a cut nail, with a rather longer head formed so as to have a spike like a little nail at one end, this being at an angle, so that as it enters towards the end of the nail's travel into the first board, it picks up and forces the next board up tight. It would also ensure the correct position for the nail at the edge of the board, and save half the work.

A TURRET DRILL

How often have you drilled a hole, and then had to take the drill bit out to insert the screwdriver bit to put the screw in? Then you probably had to repeat the process. Tradespeople usually have two drills on the go, but for the home handyman this is a bit of an expense. A turret lathe has a head which can be rotated so as to bring different tools into play without having to take them out and insert another.

You can invent the turret drill. Having drilled the hole, the drill swings sideways to present the screwdriver bit. Much time is saved and you won't lose things by putting them down after use. How many times have you mislaid the chuck key? Or found it has slid all the way down the cable by your feet? Make it a fixture; swing a lever and it engages, flip back and it houses out of the way. I will leave the details to you.

A TOOTHPASTE DISPENSER

Toothpaste is sold in tubes; a better way would be to have a dispenser. This would consist of a wall bracket, into which drops a cylinder, not dissimilar to the ones you buy caulk or bathroom sealant in, but of a different bore, so you don't brush your teeth with bath seal!

This would screw or clip into a dispenser, consisting of a plunger which, when activated by pushing the toothbrush into a slide, deposits a ribbon of paste along the brush. A return spring reloads the dispenser ready for the next use. The top of the cylinder has a free piston, similar to your tube of bathroom sealant, and air pressure pushes this down.

One advantage of such a device would be that it can be used one handed, always a desirable feature, and also a measured quantity of toothpaste would be delivered. You could do the toothpaste as well. If you do, use chalk dust as the polishing agent. I am a great believer in calcium carbonate deposited in crevices that may harbour acid forming bacteria. I know other makers use ceramic dust, but this is indigestible, has no antacid properties, and unless you have good quality control, may lead to nanoparticles getting into the body, with unknown consequences. Chalk dust will be immediately dissolved by the hydrochloric acid in the stomach, which adds to the calcium intake, not a bad thing.

IMPROVED CANDLES

It is instructive to examine how old inventions have been improved; for example the umbrella had been around for a hundred years or more before the folding variety was introduced, a very useful improvement. Again, the bicycle was a standard item for over a century before Moulton developed the small wheeled folding bike.

Other examples are the humble boot polish tin. Before the war they had a little turnbuckle riveted on the side, which when turned, forced the lid up. During the war this was abandoned to save metal, although I don't think very much metal was saved. Later, the oyster tin was developed; this had a curved cut out on the vertical side of the lid. When pressed, the tin popped open like an oyster. This saved metal and meant that the can could be opened one handed, something to be sought after. Of course much less boot polish is used now; quite often inventions come a bit late.

The candle has been around since prehistory, and one might think there would be little room for improvement. Not so. If you see a Victorian candlestick it will often have a small pair of scissors attached. This was for trimming the wick, which did not curve over as it does in present candles; that invention was post gas lighting. A single linen thread was incorporated in the wick; this contracted on burning to pull the wick over in a curve, which presented the end to the outer oxidising part of the flame, where it burnt away. You will see that it glows red just where it pokes out to the edge of the flame where there is plenty of oxygen. The centre part of the flame does not, causing the wax to burn off the hydrogen first, leaving carbon particles which glow white as they burn towards the edge of the flame. You may notice that the lower part is translucent, consisting of unburnt gas, and next a blue bit as the hydrogen burns off.

When blown out, the end of the wick glows, giving rise to a column of unburnt gas which condenses and smells. You could try soaking the wick in a higher melting point wax to obviate this problem. A Victorian candlestick often has a snuffer or extinguisher in the form of a cone, which could be held over the top of the candle. You may see them used on tall church candles.

One advantage of a candle is that it can be stuck down anywhere with a few drops of wax; this also presents a danger, for if stuck on an inflammable surface, it may set it alight when it burns right down. What to do to prevent this? Well, some candles have the wick ending short of the bottom, others have the last bit of the wick enclosed in a metal clip, or treated in some way that it will not burn, so the candle goes out before reaching the very end.

Candles often drip, which is a waste of wax and makes a mess, apart from increasing the chance of a fire when it burns right down. Some candles have longitudinal holes in them; wax runs down and prevents a puddle at the top which may spill over and cause a run down the outside. The holes fill of course, but then melt open again as the candle burns down. You could try different sized holes to keep re-melting and re-opening out of step. I had some candles that had a thin layer of gold paint on the outside. This crumpled as they burnt, forming a low rim that prevented runs down the outside. You could invent a candle that had an outer casing of harder wax with a higher melting point; this could be done by dipping or swilling the mould before filling. One minor disadvantage with a candle is that when first lit, it burns up brightly, then, as the wax burns off the wick, it dips down to a low flame until a fresh supply of wax has melted, making it liable to draughts. You could dip the wick in some low melting point wax to obviate this problem.

If you look at a candle while it is burning, you will see that the flame is not quite centred over the wax, due to the curve in the wick. This can lead to runs, which are undesirable as they waste wax and make a mess to clean up. If the wick were twisted into a spiral, the overhang of the flame would also spiral round as the candle burnt down, distributing the hot spot and reducing runs. Alternatively, the wick could be in a spiral, and twisted so that the overhang was always at the centre of the candle to give the same effect. You could of course have an eccentric wick, but this would look a bit odd and maybe they might not sell as well.

Candles need candle-sticks; a removable top that can be easily cleaned is of use. Also, as there are several sizes of candle and several of candle-stick, a mismatch often occurs. This can be offset by having a tapered base to the candle, with easily crushable ridges, so it can be forced into the stick Alternatively, three slim knives, tapering towards the top, could be inserted into the top of the stick, so the candle can be forced onto them. One candle gives a lonely light, so they are commonly used in candelabras. One problem associated with several candles together is that the rising hot air column from a lower candle can draw in the air column from a higher one, causing the flame to gutter. You could work out the minimum distance apart against height on a graph for candelabra makers.

One thing which seems not to have been developed by the Victorians, is a tube to blow a candle out. You could invent one a bit like a peashooter and work out the best length, bore, lip plate and end flattening to give a spread across the flame. Or you could invent a popgun device, where putting the trigger compressed some air which was released as a puff at the end of travel.

I am not actually suggesting that you go big time into candle technology, although you could; my point is that long standing inventions are often capable of improvement, and the statement that it has all been done before should not put you off.

One advantage of the oil lamp over the candle is that if the wick is turned broadside on you get more light on your work. It would be possible to designing an oval candle with two wicks, with the wicks arranged so that the linen thread is outwards, causing them to bend out to give a broader flame. 'Art' candle makers could do with a bit more science. They usually have no knowledge of the correct wick to thickness ratio for a candle, with the consequence that they burn hollow, give less light and end up being thrown away with only half the wax being burnt. If you are a candle maker you could explore different colours, for while coloured wax has been about a long time, nobody seems to have explored coloured flames. Ideal as party candles, a few grains of magnesium dust in the wick will cause the candle to reignite from a smoking wick. Try soaking the wick in a common salt solution; you should get a sodium yellow flame, copper would give blue to green while calcium, red. Do experiments to get the best concentration; try mixing colours. Check to see if there are any black marks against some chemicals, Strontium is a bit poisonous. You may have to use volatile compounds for some elements, to get it into the flame; cerium, calcium and thorium spring to mind. You could have a candle that changed colour as it burnt. You could also experiment with wicks; can you increase or decrease the curvature? If so, are there any advantages? Try using different materials for the wick; plastics might melt to give the same effect, a curvature, or a hollow plastic tube might hold more colouring material. It might be possible to stop the wick smoking after it had been blown out. You could take lessons from a seaside stick of rock maker, and put writing or patterns in your candles, you could put hour markers and make a candle clock which counted down showing the hours left to burn or those still to elapse; let the imagination run riot! All this from a stick of wax with a bit of string through it.

Early gas jets had a fishtail burner to spread the flame. They were still in use on the stairs of the King's Theatre, Portsmouth, when I was young. I had a cottage that still had a gas lamp left in the living room. This was quite useful for watching the old dark screen televisions, as it could be turned down.

The Wellsbach gas mantle was a great invention, an improvement on limelight, which needed oxygen and hydrogen to work well. Cerium and thorium salts were used on a cotton net cased in collodin. When lit, a skeleton of metal oxide was left which glowed white in the flame. Thorium is mildly radioactive however, so the late Victorians probably had more radio-activity in their living room than today. Cigarette lighter flints also use thorium, so perhaps the shortfall was made up as cigarettes came into more common use. An estate agent friend once showed me a house where they had never had the gas or electricity put in, using oil lamps instead; and another that only had gas.

Gaslights were the most common form of street-lighting, and a lamplighter used to come round at dusk and light them. They were of the gas mantle variety, and lighting was by means of a counterbalanced chain, like a pair of scales, with a triangle on one end and a circle on the other, a small pilot being on all the time. Later clocks were fitted that only had to be wound once a week. There were still three gas lamps left up the top of Luccombe Road on the Isle of Wight, as the gas went up the road, but the electricity didn't. I expect there are one or two left over the country still.

BETTER CRANKS

I once invented a better crank; my friends told me I didn't need to, but this one was for a bicycle. More efficient pedalling can be obtained if one can get through the top and bottom 'dead centre'. The problem is that muscles work best at certain speeds, and at top dead centre one can push all one likes and get no propulsive effort, although the muscle is doing work inasmuch as it is in contraction.

By having an elliptical chain wheel, top dead centre can be got though more quickly, further down the stroke; where the muscle is working more optimally, the stroke slows. My patent agent said it had been done before, so I did not take it much further, although I had never seen one. Later one of the Japanese bike manufacturers took it up; if you want to have a go, they do not have a monopoly. While on the subject, it seems that all cranks are the same length, except for children's bikes. There is a good case for having it adjustable for leg length; after all, saddles are adjustable, so why not cranks?

Another possible device might be one which stores a bit of energy on the downhill to be used up again on the uphill. It would need to be light, but need not store much energy. There are several possible routes; clockwork, compressed air, flywheel. You could see if any of these could be adapted.

For piston engines, one of the limitations is piston speed; go a bit too fast and you get accelerated bore wear. I will leave you to design the automotive variant!

A CRANKLESS ENGINE

I designed a crankless piston engine once. To explain this, imagine that you trace out the movement of the piston onto a moving strip of paper: You will get a sine wave. Now make this in a metal circle, and have your piston going up and down following the curve. You won't need a big or little end either; a stout rod prevented from lateral movement on the base of the piston will be all that is required. You will need curved roller bearings to take the lateral load, for as the connecting rod in a normal engine moves from side to side a lateral thrust is produced. For this reason the skirt of a piston is sometimes longer on one axis to distribute this load, so you can have shorter pistons.

Also of course you can throw away the crankshaft, which is heavy, expensive to machine and may flex slightly. The cylinder would be static, the curved ring rotating, being driven by the motion of the connecting rod along the curve, and drive being taken off at either end, or from the ring via gearing. The end of the rod would have a stout roller or wheel so that it rolled freely along the curve. To give downward movement when the piston was not firing, a spring could be incorporated, or a preventer, being a piece of metal over the top of the roller on one side to keep it down; this would also need a roller bearing round it, the movement of the top of the wheel being contra to that of the bottom.

You now have the advantage of being able to alter the piston movement rate by altering the curve; you might for example want a quick getaway to prevent pinking in a high compression ratio engine, and a quick reversal on the exhaust. This would slow the piston at the centre of its motion, thus giving a faster revving engine without exceeding a safe piston speed. Also, you could put another piston back to back with the first, or use four. A six would be a compact arrangement; for any multiple the opposed pistons could be stepped, or, if paired and firing together, there would be little net force on the ring, which could be much lighter. For very large engines, blocks of pistons could be put in line down the shaft. Valve drive and ignition timing could be taken off the same ring, or one that is capable of advance or retard, running round the outside of the main drive ring.

Thus we have a more compact, lightweight, and faster running engine. You will sell several hundred million.

AN IMPROVED SAW BENCH

Log stoves are becoming more fashionable. Often they are just that; fashion items, standing squat and black with a wicker laundry basket of logs alongside in the hearth. It gives that cottage feel. Maybe the owners try them once and get fed up with clearing out the ash. Of course, you may burn solid fuel: many stoves will burn both. This produces much more ash. I think they put extra ash in to bump the price up, or maybe it is a way of getting rid of low value coal. A simple hod stand may be made with a semi-circular hole cut in a plank, mounted vertically on a stand; this obviates the need for a coal scuttle. Hods are usually conical in shape with an angled cut off at the top, which is not ideal for tipping into a front loading stove. A better shape would be to make the hod bellied, so that as one tips the angle increases for the last bits to come out. Also, if people buy their logs they will find that wood has only about half the heat value of coal, and takes up twice the space. Often it is more expensive than coal for the same amount of heat.

For those people who can get their own wood, pallets are usually free, as the builders' merchant has to pay to take them to the tip. They may even deliver them free if you are near. The local tree surgeon may give away unwanted off-cuts. If you live near a beach there is usually driftwood, also other treasures. This leads to the problem of sawing up. The average saw bench is awkward to use; quite often they are too high, particularly if a big log is put on top. Also, the last cut can be wobbly as there is only a short length to put your foot on.

You could develop the improved saw-bench. This would have a pedal that clamped the log down on the bench, with a spring to lift it up as you moved the log down. Also, the bench could be split; it would be essentially two short benches joined at the base, with a slot halfway along so that one could pass the saw through when sawing short logs, which would be supported at each end and not wobble. An adjustable stop could ensure all logs are the same length, while the deluxe model could be height adjustable too. If you buy a saw bench, or more particularly a Black and Decker workbench or look-alike, you may find that it is too light and tends to move under load. Design a stout canvas bag with a good Velcro fastener that can be loaded with shingle or anything, and draped over the bottom cross strut. That should keep things in place! Thus an item that has been around for probably several thousand years can still be improved.

You will also need a chopping block. You will probably get some large rounds from a tree trunk at one time or another, knotty bits from where the branches come off are best as they are not so likely to split. Different people prefer different heights; if it is all too short you can put one on top of another. The bottom one is best flat, but, surprisingly, a slight angle to a chopping block can be an advantage, the reason being that some of the rounds you chop will not be square cut, so you can use the slope of the chopping block in opposition. Inexperienced axemen often miss the log, the impact being taken by the top of the shaft, which will eventually lead to cracking and the head falling off. This can be prevented by a metal tang, say six inches long, running down the impact site; it could form part of the wedge used to expand the top of the shaft to fix it firmly on the head.

While on the subject of axes, you will probably have a small chopping axe. These are very convenient for kindling, but sometimes you will wish it had a bit more weight. A clip-on weight that goes over the back of the head and handle will add this and is easily removable when not needed.

A COUNTERSINK DEPTH-SETTER

Another simple invention is a countersink depth-setter. This consists of a piece of metal with a hole through it to take a drill-bit; it can be clamped onto the drill with a grub-screw. Used one way up a flat face is presented as a depth setter, the other way it is a counter sink as well. This idea of mine was pirated in China years ago. It might be worth re-issuing though, as I have not seen them about in hardware stores.

BETTER PILLS

Do you make pills? Well, they usually roll when tipped out of the bottle. Make them with a tapered edge, as a section of a cone rather than a cylinder; they will at least roll round in circles then. Better, put a cut-out in one side so they fall over within one revolution. Or make them square or triangular, though they might then be difficult to swallow!

BETTER FLYPAPER

Do you make flypaper? Well, it doesn't attract flies, although they may sit on it occasionally. Far better to find out what smells flies actually like, then put it in the paper. If there are several, you could have breakfast, lunch and dinner down the strip. Print a couple of flies on it, just to give them the idea.

ADDED NOISE!

Usually noise is a disadvantage in a product, unless it tells you something. I once had a battery operated razor with a very quiet motor and a noisy foil. One could hear when it was cutting and this was a great advantage. Thus, if you are designing razors, you could suppress the motor noise, perhaps by feeding in 'anti-noise', that is, the sound wave recorded and played back out of phase, or amplify the noise of the cutter.

While on razors, they have not changed in at least sixty years. A vibrating rack of metal plates chops the hair off as it protrudes through the foil. An angled cutting edge would pull the hair down slightly as it chops; also, it could wear so as to be self-sharpening. Foils are a nuisance when they fail, but to last a long time they must be of a certain thickness. An optimal, thinner, foil could be used if it came in a roll embraced within the razor; when it failed a new section could be wound on with a marker like the old roll films, to show when you were getting near the end.

Rechargeable batteries in razors should be replaceable; most types can only be used as a corded razor after the battery fails, which it does long before the rest of the razor. The latest type has a little charging stand, so the razor has to be thrown away. This is of course deliberate design, however you can steal an advantage by pointing out that your razor does not have to be thrown away.

There are probably lots of examples of added noise you can think of. Subtracted noise could also be used. How often have you seen a programme spoiled by inappropriate intrusive and overloud music? The ability to remove the music would be a great advantage, and this could be done by briefly sending it to a noise extractor, which took the music out and sent it back an instant later. Orchestral instrument sounds could be easily separated from speech, or at least muted. There is an app that would make money.

All mechanical items should come with a decibel rating. A friend once bought a much advertised vacuum cleaner; she said it was incredibly noisy.

IMPROVED AIR QUALITY ON PLANES

Planes are great spreaders of disease. You are shut up in a box with three hundred other people for hours on end, and you all have to re-breathe the same air. I once sat next to a woman who coughed all over me for eight hours. They should have special seats reserved for people who cough, next to the air exit, or chuck them out instead. It was much better when people were allowed to smoke; they had to change the air more often then. Most passengers have had a bad air trip at one time or another. The problem is that it cost money to take air in at thirty thousand feet, compress it to the equivalent of eight thousand feet, then throw it out again. Mitchell, when he was designing the Spitfire, was worried about the drag occasioned by the air scoop underneath that scooped in lots of air to cool the engine. He need not have worried; the air expanded as it warmed up, and actually produced a few pounds of thrust. If you design planes, use the same principle; the warmed air from the cabin could be vented so as to produce thrust to off-set the cost of compressing it, then we can all breathe again. The problem is compounded by the fact that compressed air is drawn from the primary compressor for the engine, and this is always polluted to a greater or lesser extent by lubricant, principally phosphate oils. The answer is to have a separate compressor, which could be greatly assisted by a ram-air effect; that is, a scoop to compress the air using the forward speed of the plane, which is what they used to do in the old days. Filtration would remove aerosols, but not volatiles, in the interim. The industry is well aware of the problem, but keeps quiet. One way to force the issue would be to produce a simple hand held detector to identify the worst pollutants and oxygen levels, which you could sell at airports until they banned you so that you had to set up your pitch outside. You would sell millions. Once passengers were aware of the risks, and started demanding compensation, there would be a rapid shift in the industry.

A GOD IN YOUR POCKET

Long before Ishtar (aka Venus), Artemis, Diana (the Roman Artemis) Isis, the Queen of Heaven and Mary, also known as the Queen of Heaven, in the time when there were only six signs in the zodiac (if you ignore the serpent, which wandered through all,) we were ruled by the moon, our mother who dragged us onto land across the wide littoral. Deities were then female, and the earliest was the fat lady, a fat or pregnant highly stylised statue of a woman, with small head and small pointed feet. The origin seems to be in that vast, little known (at least to Westerners) region to the north and east of the Black Sea; a grain basket and cradle of civilisation, where east meets west, or maybe where both were modified. The fat lady has been found all round the Mediterranean, a symbol of plenty and fruitfulness.

The nearest thing we ever had, but on a much lesser scale, was 'Joan the Wad,' the Cornish Pisky, models of which used to be free for a shilling. I expect they are free for a pound now. Worry beads were in fashion at one time, and Joan the Wad performed a similar function. You could make a small plastic model of the fat lady and sell it as a good luck charm. With the decline of religion worldwide there should be good sales, for as one Anglican Bishop remarked in the nineteenth century;

"The problem is that when people cease to believe in God, they do not believe in nothing; they believe anything!"

It might be advisable to keep in her good books, for she too will have noticed the changes we have wrought in the environment, and, before long she may be getting ready to sing.

RECYCLED BRICKS

Have you ever passed a demolition site? It is amazing how much good stuff goes to landfill. Bricks are a case in point; old mortared bricks are not a problem, and the mortar usually comes off cleanly with a few bings from a trowel. Why bings? Because that is the noise it makes, even if your spell-checker does not like it. The problem comes with cement; it can be extremely difficult or impossible to get off. What is needed is a device with two cutting wheels; you slide the brick in, and it takes the mortar off two faces, then it passes on to another two to clean up two more, then again. If the brick is firmly clamped, you could try two cutting edges that descend instead. You could sit there all day, putting a brick a second in; they may be worth fifty pence each, so thirty pounds a minute, or one thousand eight hundred pounds an hour; better than a dentist! Even at ten pence each it would be worth doing. The sand could be recycled as well. In fact all of a building should be recyclable.

I once made some model bricks, using chalk. This has been used from the distant past. The method was to fix two stout planks between uprights, then pour in crushed chalk, which was damped and rammed. Then another set of planks added until the desired height reached. The building thus constructed had good insulation properties and used the local natural material. Mud walls were sometimes made the same way. In France they are still to be seen, usually as garden or field boundaries; quite often they have a tile cap to shed the rain, and may be plastered.

My chalk bricks were made of chalk powder, but mixed with five per cent pitch, which can be powdered when cold. This was then subjected to high pressure in a vice and heated. The resultant material was hard, waterproof, fireproof and looked a bit like limestone. A plant in a chalk-pit could produce a high quality building material, better than concrete blocks, cheaper and with no drying time.

INCLUSIONS IN PERSPEX

I once had the idea of producing a decorative material by

including other reflective materials in Perspex; crystals, wire turnings, stamped designs such as stars and diamonds are some examples. I wrote to I.C.I.; they patented it in their own name immediately. They never produced it though; it was merely a blocking move. Their patent will have expired by now though, so feel free to have a go.

Another example of dodgy tactics was when I was working on a safer cigarette. The idea was to smoke the tobacco, remove the carcinogens; then re-absorb the residuum onto carbon or some other suitable material.

The tobacco lobby got the law changed so that any such product would carry tobacco tax. They were hoist with their own petard however when they later produced 'new smoking material', essentially the same idea using cellulose as the base.

BETTER SAUCERS

I once made a saucer with three small ridges on the inside, where the cup sits; the idea was to keep the base of the cup out of any spills, so that it did not drip when picked up. This is a pretty obvious idea which has probably been done before.

TEA BAG TONGS

Tea bags are very useful, but they tend to drip. A pair of tongs to remove them from the cup and give them a squeeze would be handy. You could design a pretty pair that would double as sugar tongs, or a cheap plastic pair to go with every big pack of tea bags

BETTER PLOUGHS

Ploughing take a lot of energy, and this means a heavy vehicle that can compact the soil. I once designed a device like a harrow but with hollow tines. Compressed air was blasted through in separate little explosions, to break up the soil. Fertiliser and/or seed could be added at the same time. The advantage being that the energy to move the soil was put in by the air, not via the towbar, thus obviating the need for a heavy tractor.

BETTER FILE REINFORCEMENT RINGS

Ring binder and string files are still in use. A problem arises when the paper tears however; the file has to opened at the page, a file reinforcement ring stuck on, and the page reinserted. For a string binder file you may do as much cumulative damage to the rest of the pages in the process.

A simpler reinforcement can be made by kiss cutting a string of 'U's, half of them inverted so they nest into each other. This also saves material; in cutting the usual ring, half of this is thrown away by the hole in the centre, and the surround of the ring when it is cut out of the sheet.

The new style reinforcement does not require the folder to be opened; the correct place is found, and the 'U' inserted with the open end facing across the page, the rest of the 'U' taking the place of the torn section. The 'U's can also be used as page index markers along the top or side of the page: two are used, back to back, to make a tag to mark the page position. My wife won the Perstorp competition with this one, many moons ago. A similar invention, that of cutting without wasting, was;

DOLLS FOR GUYS

The guys in question are tent guys; they can be of thin plastic and children run into them, or you, in low light conditions. The dolls are cut-outs, with a slot from the feet leading to a 'T' cut so that they can be used in a similar way to a clothes peg, which they can also double as. They are fixed on the guy at suitable intervals making it visible. Use 'dayglo' plastic if you like. They are designed so that one doll is upside down to the next, a la Escher, so that no plastic is wasted.

A SEPARATION COLUMN

Some materials need to be separated from a large amount of unwanted dross they come with. Diamonds are a case in point. Usually they are washed out over a greased screen; they tend to stick to it while pebbles do not. This can use a lot of grease and may let the big ones through.

A better method is to use a separation column. This consists of a funnel shaped device, with water flowing in at the bottom and out at the top. Sieved material is dropped in; the speed of water flow decreases as it travels up the expanding column, and this separates out different materials against density, or rate of fall though the water.. At points down the column, there are take-offs, where water with sought after material can be extracted. Using glass or some other transparent material means one can observe what is happening and fine tune the water flow.

If you run a diamond mine, you will produce several million tons of dross for every ton of diamonds. Forget the diamonds. Find a use for this and let the diamonds be the by-product; you could explore building blocks, road fill and so on to find a market.

NON-MATERIAL INVENTION

Maybe you are interested in social invention. This usually embraces a principle, rather than a mechanical object. Votes for Women are an example, as also income tax, although not so welcome. Actually it is a bad tax, as it may discourage effort from high earners, who are often highly productive. The old purchase tax was much better, although now euphemistically called "Value Added Tax". There is no value added, only tax. The principle advantage is that it is infinitely tuneable; highly resource wasteful or dangerous products can be heavily taxed, while others not taxed at all. There might even be a case for negative tax for such items as bicycles or energy saving electrical appliances.

To give some specific examples; how about legislation to the effect that any hydrocarbon fuelled road vehicle should not exceed the weight of the load it is intended to carry? Buses, lorries and most motorcycles would be OK, but the average car, not. An incentive to manufacturers. Electrical transport could be exempt, as some use lead acid batteries. The idea should be presented as a goal to be worked towards, by stages, so as to allow manufacturers to catch up.

Again, how about legislation for supermarkets? Whenever you go into one you are bombarded with "Buy one get one free". An obvious lie, for if you pay your money and get two, it is two for the price of two, with the added penalty that if you only want one, you still have to pay for two. 'Massive price reductions,' when the price was previously raised for a few weeks before. How about legislation to the effect that any price increase should be given equal advertising to that used for reductions? 'One for the price of two!' 'Massive price increases!' 'Don't hurry; we are going to drop prices next week!' This would be more honest, but might not look nearly as good to the management.

Another social invention could be a tax on undesirable television content. Everyone outside the industry knows that violence is learned, and the observation of violence is likely to generate more. The industry knows this as well, but chooses to pretend not to. There are well documented papers available to illustrate the principle. How about a gun tax? Every shot filmed could carry a tax of one tenth of a penny per viewer, with killings at a penny. While we are at it, a smoking tax could also be applied when smoking is shown; after all, the film company was paid to put brands in, so why not negate this? The tax collected could be used to offset the licence fee, or be returned to the industry to make better quality products.

While on the subject of guns, there is a problem in America and other countries with the widespread dissemination of guns amongst the population. America has a very high gun death problem, while any attempt at restriction of use is strongly fought by the National Rifle Association, which quotes the second amendment, which allows people to bear arms. Different states have different interpretations, but generally you can go into a gun shop and buy a gun with little difficulty. The NRA of course is not so much protecting the citizen as protecting the manufacturers of weapons. We have here an example of a business which is pursuing profit, a reasonable objective, but to the detriment of society at large. The primary goal of any business is to pay shareholders, and those who run it, not to do good to all and sundry, however much they might protest otherwise. Many businesses, the tobacco industry, wines and spirits, and the drug companies (which are the squeaky clean counterpart of the two previous examples), all to a degree engage in practices which are questionable.

One simple remedy might be to pass legislation to the effect that all guns and ammunition should carry insurance. We have an example in motor insurance.

There could be provisos; for example, if you leave your car unlocked with the keys in the ignition, some insurers will not pay out if it is stolen. Similarly, if you leave your gun not secured by lock and key kept in your possession, the owner could be made to pay the cost of any damage done by somebody taking it. This could be very expensive if a neighbour's child shoots themselves with it, or the unstable lad from over the way shoots a dozen on campus. Once you are bankrupt the insurance could pay out any costs not covered, so as to protect the general public in some measure. There could be a distinction made between concealable weapons, such as pistols or revolvers, which are much more likely to be used in crime. After all, although the second amendment gives you the right to bear arms, if you wandered about New York sporting a couple of hand-grenades on your belt, with perhaps an anti-personnel mine and some nerve gas as a makeweight, you might find the police raised an objection, second amendment notwithstanding. It is all a matter of degree. In the UK concealable weapons are banned.

Presumably, if someone buys ammunition they intend to fire it off, but quite often a box or part box is left over; after many decades it might turn up in the shed or loft when you sell your house, to be found by young children or put on the bonfire long after you have forgotten about it. Ammunition should have a 'best before' date, say a year after manufacture, after which it degrades and will no longer work. You would not want to use a loaf or bottle of milk that is a year old, and ammunition could be a similar case. Manufactures would welcome the move, as it would give a continual steady state re-supply and reduce stockholding.

The NRA would of course object, but it would lead to great increase and attention to safety. The added cost would be objected to, but if it cost the industry and gun holders half a billion if someone went on the rampage in a primary school, rather than expect the bill to be picked up by society at large, their remedy would be to make this a far less likely probability, to everyone's advantage. Many industries become dysfunctional. The drugs industry is another: the ambition is to have everyone over forty on permanent medication; to find no cures, merely to have endemic, treatable diseases. They do their own testing, with no independent review, and fudge results. A good proportion of the advertising budget goes on straight bribes.

Some proposals could stir up quite a controversy. How about a net zero cost prison industry? It should not cost more to send a person to prison than University. After all, prisons do not have to spend much on children or old age pensioners. Workshops or farms producing useful goods could be set up, with the prisoners getting a small proportion of the proceeds and the rest going to running the establishment. Outside firms could licence work inwards, as they would already have the markets established. Given a population of mostly healthy youngish men, productivity could be high and pay for the running of the prison. Inmates would receive useful training for when they rejoin society.

I expect huge opposition would come from prison staff, and do-gooders would cry "slave labour!" This all adds to the fun though; there was opposition to votes for women, and the factory acts that banned child labour, but they are accepted now.

The biggest support might come from the prisoners themselves, particularly if there were perks involved. Once familiar with the system prisoners could be promoted to run it, with a little oversight; the best could be allowed home for the nights, with living out allowance, and great saving on accommodation costs. New accommodation could be built by others, at a higher standard than that existing, assuming that we will then need it of course; at least we could knock down some of the old stuff.

IMPROVEMENTS TO PENSIONS

Most people are beginning to worry about their pension, or if not, they should be.

The problem with the State pension is that there is no pension fund. When set up, the idea was that taxation for those in work should provide a pension for those who had retired. At that time the ratio was about one to ten; now, as people are living longer and the birth rate is down, it is about one to four, and people draw their pensions for longer also. Despite raising the pension age for women from sixty to sixty five, and later, raising the retirement age to sixty six, then sixty seven, the system will eventually fail to provide a reasonable pension. The answer is to establish a pension fund; the problem is; where from?

About the only viable source is the stock exchange. Trillions are transacted each year, and the traders take a small percentage on each trade. As some of these trades are multi million pound deals, they get a large return for very little work. It is unreasonable to have to pay thousands of pounds for less time than it takes to buy a bag of sugar at your local shop. Thus deals could carry a small tax, payable in shares, which would form the basis of a pension fund. Initially, the incremental earnings of these shares could be re-invested in the scheme, and the shares could be more or less static, with little churning; if you have a wide spread over the market, then you are the market, and churning does not work well, unless you go in for manipulation, that is. Most shares generate far more than their original cost by traders churning and taking a slice each time they do so. One could also look at company tax; this could be paid in shares also. This would be of advantage to the company; they could have a year in which to pay, thus buying in shares at what they consider the lowest price, while handing over at the peak, or they could print new shares, thus diluting the stock.

Eventually, there would be sufficient income from earnings on the shares to provide a pension, tapering in the while. One objection might be that in times of slump, pensions would also slump. There is nothing wrong with this. Everybody would take the strain, while there would be good incentives for employees to maximise the performance of their company, as their pension would be tied up in its success.

Another route might be to tax companies that use dishonest advertising. Although it is reasonable for a company to put the best face on their product, downright lies should be discouraged. For example you may see tobacco advertised as Cool, Rich, Satisfying. On closer examination we find that tobacco burns at somewhat over 500 degrees centigrade, hardly cool. Then exactly who gets rich? Not you. If alternatively invested in a pension scheme, the smoker would get an additional pension similar to the old age pension. Satisfying? Smoke one of these and you will want another in half an hour or so. Dissatisfying might be more appropriate.

Suppose we had a scheme where, if dishonest advertising were proved, a tax equivalent to a fraction of the total share value of the company were levied, to be paid either in cash or by a share float, which could go towards the old age pension scheme. Advertisers would be a lot more careful, and might get adverts vetted first before publication.

Another issue to be addressed is the responsibility for unsafe products. If damage is caused by an unsafe product, they there should be a liability on the producer to rectify. Thus if tobacco is shown to cause premature death, then there should be some form of compensation as a matter of course rather than each case having to be separately fought. It would encourage safety. The tobacco industry has been lax in this area, to their own detriment: if a smoker lives X years less than a non-smoker, then the company has lost X years of sales.

DID YOU HEAR A HORSE?

I have heard that the four horsemen are on the loose, so perhaps we should be prepared. At the time of writing, bubonic plague is in China, caught from a hamster, no less! This was not so very unusual in times past. We had some old candlesticks, from my grandfathers' antique shop, when I was young. There was some yellowing German newspaper, in Gothic print, as part of the wrapping; we had it translated and it was about the plague in Manchuria in the late nineteenth century. The Black Death across Europe is thought to have been bubonic plague, but it may have been a look-alike or some virulent variant. There is also Ebola, not under control in Africa, which could be much more serious. Of course this will all be history by the time you read this, or maybe we will. If it all fizzles out, don't worry! There will be another one later.

There are only two things you need to know about an emergent disease:

1) The survival rate, and

2) The doubling time.

It is also useful to know the time for which a person is capable of infecting others.

The survival rate for Ebola is ten per cent, and the doubling rate roughly a fortnight, which may decrease. At the current death total this equate to something over a million at the end of 2014. Another year brings us to a figure of about half the world's population.

So what should we do? Firstly prepare a plan, for sooner or later a major epidemic will occur. All movement of people from an infected country should be stopped. Then the first line of defence is a stage one lockdown; this means that where an infection arrives by plane, train or boat the surrounding area is sealed off. Nobody leaves, or if they have to, quarantine areas should be set up, this as much as to reduce the probability of people fleeing as anything else. Quarantine areas should be small, for if anyone proves infected, then everybody else is back at square one. If anyone wishes to enter, they stay there. We could look at items leaving also, and if we cannot be sure they are sterile, they stay inside the containment area too. Some will protest about disruption to trade, jobs lost and so on, the quick answer is 'so what?' Containment overrides all. Better to have no trade and live people rather than no trade and no people. Councils within the area could be given broad ranging emergency powers so they are not inhibited by threats of legal action, breach of human rights and so on.

Quite often initial containment does not work, so we will then go on to stage two lockdown. Here exclusion areas are set up where nobody is allowed in. The same thing happened in the Black Death: villages sealed themselves off against people fleeing the towns; trade was continued by dropping coins in vinegar at exchange sites, to limit risk. Airports and docks should be closed to passenger travel, while goods should be checked and sterilised as far as possible.

The wearing of masks should be mandatory in all public places, and assemblies, football matches, church gatherings and the like banned. A stockpile of several billion masks should be available for immediate distribution to surgeries and councils. People failing to wear a mask should be fined £50 and immediately transported to a 'sin bin.' It should be explained to them that their mask will prevent them being infected by all these other people who are at enhanced risk because they don't wear one. Once they have understood, they could be let out again, with mask.

An exercise in doomsday scenario writing perhaps? Maybe, but it would be better to have a plan ready rather than be scratching about at the last minute.

SOME ONE LINERS OR NEARLY

Quite often an invention can be expressed in a few words; 'Put a cart on rails' was the original idea behind the railway, for example. Of course, there is a lot of expansion of detail, materials testing, and measurement and so on before you have a working model.

Here I will give some outline ideas, and leave you to do all the hard work!

1) Aeroplanes sometimes have to land with the undercarriage up, due usually to hydraulic failure. They can get down in one piece by skidding along the runway, but this leads to clouds of sparks, which can cause fire.

The problem is that aluminium or its alloys with magnesium is highly inflammable once the oxide coating is scraped off and small particles or droplets exposed to the air. This could be alleviated by putting small skids made of phosphor bronze or gun metal on the bearing points, as this does not give sparks. Flooding the runway, particularly the far end where the plane comes to rest would be a good idea also. Farm irrigation gear could be easily adapted.

2) While on the subject of fires, oil tanks are particularly difficult to put out, as water pumped in sinks to the bottom, and foam can be destroyed by the high temperature at the top.

This problem could be solved by pumping in a number of small hollow steel spheres, ping pong ball size or less. They could go through an impeller type pump along with water or be air blown, and would float on top without being destroyed by the temperature; a kind of indestructible steel foam a few feet thick. Early test pieces could be made from a thin walled steel pipe cut every inch or two, the cut sealing the edges and leaving a short sausage, although these might not pump well. Water then sprayed on would cool these and reduce the temperature, or condense the gas from the vaporised oil, leading to rapid extinction of the fire. Again, a farmers' irrigation system to spray water across the top and down the sides of the tank would be cheap to fit. The neighbouring tanks could be similarly sprayed to prevent them catching fire also.

3) More about fires! You can run your car on whisky, although it is only about 40% alcohol I have heard that someone has run a jet engine on 15%. Such an engine, particularly if water was sprayed into the exhaust, would produce a large quantity of 'vitiated air' (most of the oxygen removed) and wet steam. Directed at the base of a fire this would cool the burning material and remove oxygen, thus putting out the fire.

4) Have you noticed the proliferation of LEDs? They come in all kinds of colours now; it seems every extension lead, computer, switch and mousetrap has to have at least two. I have twenty-four, including a couple I discovered the other day on the back of my old printer, which I have had for a decade. I know they don't take much electricity to run, but most of them are on twenty-four hours, three hundred and sixty five and a quarter days a year. I don't know the lifetime of the average appliance, but call it ten years, then multiply for the number of houses in the world that have them, or are likely to, and you have several large power stations burning hundreds of tons of coal or oil an hour. You could start a campaign, 'Lose the LEDs', or at least have them switchable, and not with another LED to tell you they are off. Just think; you could save tons of fuel per second.

5) Bartenders can spend a lot of time pouring bottled beer slowly to prevent foaming, while customers are kept waiting; a good bartender can hold two in one hand. A simple lazy spring device that the bottle can be dropped into, with the glass under, would speed things up. As the bottle emptied, it would rise, preventing a long drop that might cause the beer to foam, and re-ingesting some of the foam into the bottle.

6) As an improvement to an electric blanket, a porous double sheet, segmented like a leaky lilo, can be used in conjunction with a hair dryer. It is very fast and provided the hair dryer is temperature controlled so as not to fry hair; safer than an electric blanket.

7) Road surfaces degrade; surprisingly this is not so much due to abrasion, but oxidation. Experimentation with additives that reduce or prevent this could extend the life of a surface. Vitamin C,

surprisingly, will do the trick. However there are many other anti-oxidants you could try.

8) Cars have milometers; many have a 'trip' setting, which tells you how far you have been for a particular journey. As an alternative, a countdown variant could be used; this would be set for the journey distance, which would reduce, showing he distance left to go. This would be more convenient in route planning, as petrol gauges already 'count down' to zero.

9) One problem in gem sales is the large amount of small gems available compared to big ones. These could be used by faceting one against another, rather than separately mounting, to form a larger gem. By calculating the different refractive indexes, a change of colour for different angles could be achieved as different coloured stones refracted by different amounts. A lot of diamonds are black, being of industrial grade only; these could be used in jewellery in conjunction with other stones.

10) Many more safes are stolen than broken into on the premises, for obvious reasons of time, noise and tools. Lead weights could be sold to fill unused space, making the safe more difficult to transport.

11) Energy saving is becoming big business. How often have you walked down a corridor in an hotel or public building to find it is artificially lit, although broad daylight outside? This waste could be obviated by bringing the daylight inside via light pipes. These are filaments of glass fibre, commonly called optical cables because they supplant copper ones in transmitting data. A collector, in the form of a mirror, plane or curved, or a ball similar to that used in sunlight recorders could feed light into a sheaf of light pipes, which could then be threaded to where required in the building A diffuser in the form of a simple translucent globe could then do duty as a light bulb. The market, worldwide, could be worth a billion or two.

12) Dehydration is common in cases of intestinal infection, such as cholera, due to the inability of the infected gut to absorb water. Ebola, the next big one waiting to happen, also causes dehydration. A drip can be used, but, where this is not readily available, an alternative could be a Bernoulli tube, dipped into sterile water. By sucking, a fine mist would be inhaled. Local drinking water, boiled, could be used. If a pressurised device were needed, a simple container pumped by a bicycle pump could provide the pressure. You are never far away

from a bike pump anywhere in the world. Alternatively, use a bulb like those found on a scent spray. The lungs readily absorb water; for this reason fresh water inhalation is much more dangerous that salt, for while salt water may be got out by putting the half drowned person face down and pumping the chest, fresh water is rapidly absorbed in quantity, which may led to red cells bursting.

13) A rake is a common garden implement; a clip on shroud can convert it into a broad hoe for levelling, much better than the back of the rake.

14) Nearly all bikes have an adjustable saddle height, while virtually no scooters and mopeds do. Quite lot of models are distinctly awkward for shorter people; sales could be greatly increased by including this feature. Also, most of them are too heavy; most small two wheeled vehicles are used in towns for short journeys. The handleability and fuel consumption are greatly increased by a low weight, for short journeys in traffic involve a lot of speed changes, and acceleration costs fuel, while constant directional changes and foot-down stops need something light and nimble.

15) Cars are quiet at low speed in towns, electric vehicles more so. This presents a difficulty for pedestrians or when overtaking cyclists. Sounding the horn is inappropriate, so a noise generator could be added to the car or other vehicle, not obtrusive, but some kind of low level noise that would alert others to the proximity of the oncoming vehicle. It could be switched on when required and different kinds of vehicles could be ascribed different noise ranges. Teens already do this by playing the 'square wheels tune'; you know the one. It has lots of heavy bass that can be heard approaching before any engine noise. Bump! Bump! Bump!

16) Lighter lighthouse lenses.

Most people have a hand lens, even if they cannot find it when they want it. They don't scale very well; if you want an eight foot one for a lighthouse, it would weigh tons, so a Fresnel lens is used. If you imagine your eight foot lens, cut into concentric circles, then cut away from the back of the tubes thus formed so that they all lie flat in the same plane, you have a Fresnel lens.

We can improve on this however. If you now substitute curved rings of stainless steel to mimic the inward surfaces, you have a much lighter product. I don't know if many lighthouses are built nowadays, but it could be used for big lights which serve a number of functions. I expect this has already been done, but I have not heard of it for lighthouses.

You could even design one of several acres for a space telescope, although a thin plastic balloon, blown up with a very slight gas pressure and an outer stiffer ring like a tambourine, would be better. Don't think of the accuracy of the surface; think of the light gathering capacity. You could make a space one half a mile across. Don't fly through the focus though; it would be as hot as the sun!

AN IMPROVED GROYNE

Groynes are a common feature of the seaside. They usually consist of a line of piles, either wood or metal at right angles to the beach extending seaward, with planks of a hardwood, usually alder, greenheart or some other tropical wood, bolted on to form a low wall. Their function is to slow the longshore drift of beach material, trapping it to form a wave breaking buffer zone and thus slow erosion.

They are costly to install, as one has to 'work the tides', which are roughly an hour later each day, so that one may end up working nights or leaving gaps in the schedule. Also it is sometimes difficult to get a piledriver onto a beach, and off again, to prevent it being immersed.

Further, they don't always work that well. The problem commonly observed is that although they do trap some material at the top, the bottom end does not, principally because they break waves coming in at an angle, causing a rapid backwash down that side of the groyne which removes material as fast as it can accumulate.

So we need to improve the function and installation of groynes, the more so as with global warming we are likely to see a rise in both sea levels and surge events. Managed retreat is all very well for some areas, but in others it is not practical or desirable.

I therefore propose that we do away with piling, and use a 'bookend' design. The type of bookend I mean is the pressed metal flange type, which is not heavy enough to work by its own weight, but uses the weight of a couple of books standing on the base of the flange, which is tucked under them, to provide an anchor.

We could make our portable groyne out of an eight by ten foot sheet of half inch steel, bent at right angles in the middle, to provide a ten by four groyne, where the bit flat on the ground is anchored by beach material heaped onto it, although it could be pegged initially. I quote the size by way of example, for they could be of any desirable size, or ratio of flat base to upstand for that matter, dependant on where they were to be used. They could be bolted together to extend as far down the beach as required.

The advantage is that they merely have to be emplaced by a tractor with a bucket, which can then scoop material in to stabilise them; laying would be far faster than piledriving, and several could be laid on one tide.

Next, we must examine the failure of the seaward end to trap material. What is required is to slow the water, not stop it as does a solid wall, so our groynes should have holes in them to let some of the water through. This might let some material through also, so we have to consider hole size in relation to beach material, and the overall ratio of holes to groyne area. Eventually a kind of filter bed would be built up that would stop material smaller than the holes, by reason of the larger pieces being trapped first. If a groyne failed to work, as they sometimes do, it would be immediately evident, and the groyne could be repositioned, rather than abandoned.

Thus we have a cheap and simple system, which could be built to almost any scale, easy to position and remove, which would allow experimentation on difficult sites, and which would also be easy to repair in event of storm damage. They would stack well for transport too, and be easy to scrap when worn out or redundant. Used along the seashore instead of at right angles they could also be used as harbour walls, river bank reinforcement or quays.

HIGH LEVEL ROOF RAINWATER STORAGE

One of the strongest growing retail sectors in the past few decades has been the Garden Centre. A good garden is now a selling feature for a house, and people will spend thousands on garden furniture, pergolas, barbecues and odd looking pottery burners with a chimney of indeterminate use.

Gardens need water, which is an added cost now many properties have water meters, and now we have drought orders in April, a sign of things to come as the climate dries, evaporation and population density increases. Water butts have shown a third increase in sales so far this year, and could double by the year's end.

Herein lies a problem; the average water butt is not well suited to its function. Commonly around fifty gallons, this is insufficient for more than a week or two. Also where do you put it? Under a drain pipe of course, but this may well obstruct car access, a path round the house may be unsightly at the front of the house and just where you want to put pot herbs just outside the kitchen door at the back. Also, ground-mounted, there is insufficient head to connect a hose.

Enter the high-level storage facility. Consider a pair of tanks, one ten feet long and the other eleven (to make the corner) mounted just under the eaves. If only a foot wide and three feet deep we have three hundred and ninety gallons of storage; put the same again on the opposite corner and we have double. We have no ground footprint and sufficient head to run a hose.

The corner of a house is the best location because we will have nearly one and three quarters of a ton up there. This is no problem for the foundations, for every brick overlaps two and by the time you get to the bottom of the wall you are well round the corner. However, a load hung outboard at the top of a wall isn't a good idea, for the wall is weakest there due to the lower compression load, and the leverage greatest. The wall plate and crossties should be able to take the load, but sometimes cross ties are not much more than rested on the wall plate, with the odd nail or two.

By using a corner we can bring the centre of gravity inside the corner with one simple diagonal strut. To find where that is, merely draw a line at right angles from the middle of the two tanks; where they cross inside the loft, there is the centre of gravity. If anything moves, which it won't, the corner will be drawn in more tightly instead of spread.

We will need an intercept for the water; this can be simply done by cutting the downpipe, and inserting a crank angle to drop the water into the tank alongside the cut end, which then acts as an overflow when full. A piece of ordinary garden hose could be tucked in the angle of the pipe and the wall, and fitted with a screw or pushfit hose connector so the garden can be watered in the normal way.

How much water is available? Given an average footprint for a house, including eaves overhang of around 600 sq. feet and an average rainfall of two feet, we have a potential of around 7,500 gallons; a very useful amount. It will mostly come at the wrong time of year of course, but the keen gardener could easily fit other storage elsewhere since he can run it in by hose. A low level butt near the veg plot is not such a problem if you do not have to go to and fro with a watering can from the house each time, and a trickle irrigator could be used at low pressure.

The measurements for tank size are or course just a 'for instance' example. Any size or shape could be used, perhaps longer and shallower for a bungalow to give head clearance, or fatter and deeper for the enthusiast. A range could be developed to see where the market lies, and bolt-ons available once someone has bought and liked the product. Fitting would be easy; wall brackets or a hung bracket from the wall plate would suffice. Freezing would not be much of a problem as plastic tanks distort and return well, or a few tennis balls or similar could be placed inside. A lid for occasional cleaning and a filter grid for autumn leaves might be useful, while a tight fit would prevent mosquitoes breeding, or a spot of olive oil put in if there was a perceived problem.

So here we have a simple, low tech highly saleable product, which could be extended in range and other fitments made available; low pressure spray guns, toilet water feed, even hot water feed for high chalk districts, as a lot of people prefer rainwater for washing hair, wool, and delicates. The marketing and sales structure is already present in garden centres. It might be worth exploring 'licence of right' to current producers once patented, so long as they do not see it as too competitive to current ranges.

Development time would be minimal, as would initial cost, and time to market would be short; uptake would be rapid as it replaces or supplements an existing product that is going to probably double its sales this year. Need is accelerating and perception of need also, the double whammy effect. Also they will be visible and so self-advertising to a degree, conferring green status in the process!

INTRODUCING THE UN-PLUG

Have you ever had a sticking plug, the three pin electrical variety, which refuses to come out of its socket?

It has happened to most of us; after crawling under the desk to unplug the old printer, with the new printer plug in hand, we try to pull the plug out and the whole six way junction box comes with it, despite pressure with the heel of the hand. So what should have been a one handed job becomes a two handed one, and we have to put down the new plug to hold down the box. It still won't come loose, so we inch forward and kneel on the upturned new plug, banging our head on the desk as we jerk away.

In the kitchen it is a daily occurrence. The whisk won't come out of its socket, so we have to put it down, making a mess on the worktop, then the three way adapter comes out instead. I don't know why they are always looser than what is plugged into them, but they always are. As if trying to please, the toaster plug then drops off into the soup and what should have been a simple operation starts to spiral out of control.

Most people get used to this as a fact of life; man is not really a tool making animal, but a rubbish making animal most of the time. However if you think you have to live with this minor annoyance for the rest of your life, think again.

What you need is the Un-Plug. It is called the Un-Plug because it is not a plug, but rather a socket, designed to lubricate any plug inserted into it.

It looks very much like an ordinary socket, other than that there is no provision for an electrical connection, and it can be simply screwed onto any convenient surface if desired.

Inside there is a felt wad, cut so as to receive the prongs of the plug, and this is soaked with a lubricant. Graphite is quite good, but it is a conductor, so might eventually build up so as to cause a short, although this does not seem to happen in motor brushes. A silicon oil or grease might be best, as this is non-conducting, non-inflammable and will not break down to yield a conducting carbon residue. All you have to do is to plug the offending plug into it to give it a coating of lubricant and then replug it into the offending socket, perhaps repeating a few times to free it up. The Un-Plug can be conveniently screwed to a handy position on a workbench, or left free in a drawer for occasional use. You could even supply a little oil can, plastic of course in case anyone tries to insert it into a live socket, so the Un-Plug could be recharged via a hole on top.

It could go on sale from the electrical counter, with suitable advertising. Variants could be made for different countries. The hoover will now hop out of its socket every time you get to the end of the cord, but that is a good thing as the cord will not stand all that much pulling before the wiring tears out at one end or another. You need an extension. While buying it, get another Un-Plug for a friend as well!

While on the subject of plugs, have you ever wired one? Most people have; first, you have to get the back off, using a Phillips cross head or blade screwdriver. Next, there are two screws that tighten a piece of plastic over the cable to stop it slipping out of the plug. These are likely to need a different screwdriver from the one holding the back on. Then there are three little brass screws to hold the ends of the wire in. These will of course need a third size of driver. Usually the wire makes a few tight turns to get to the screws on top of the prongs, while you are at it, the fuse and holder will hop out onto the floor. When you finally re-assemble it, you will find that you have forgotten the cable has to go through a hole in the cover, so you will have to do it all over again.

The problem is that the plug originated as a one-off made by instrument makers in the early laboratories and then went into production unmodified. You of course will invent a proper one. This will have a back held by a simple lift up clip which also captures the cable. Then the cable will feed into another clip which, when closed, will force three prongs through the insulation and into the wires once they are separated and in position. Cables are colour coded, and the three positions will be also. No tools will be required. You will sell several hundred million. If someone beats you to it, then design a similar type socket to go on the other end of the cable

If you end up late the in the race, design a simple spool device to capture those overlong cables that trail under your desk and tangle together, or a simple piece of plastic with protruding hooks, so you can zigzag the cable to shorten it.

There are lots of other annoyances that we accept which we should not. Our devices should serve us, not the other way around.

A BETTER PILEDRIVER

"Piledriving is more of an art than a science," an architect friend said sadly to me once, after his contractor had beaten the tops off his piles, having met some hard ground. I think he was more concerned with the lack of science than the artwork thus presented, but it got me thinking.

I like piledrivers; when I was young they had steam ones, a double acting piston hefting the driver up a slide then slamming it down again on the return, ker-tok, ker-tok, ker-tok. They could do a couple of strokes a second and were infinitely adjustable for rate and strength of hit as they had a throttle. I haven't seen one for ages now; they are still the best kind though, I think, if you can get one. Nowadays they are mostly a winch with a cam-operated clutch and work by gravity, which is a bit slow. Things only fall sixteen feet in the first second, and only four in the first half second, acceleration being what it is, and then you have the wind up time as well.

So I wondered how they could be improved. It depends what you are driving, of course. I have seen all sorts of things driven, from foot-square, twenty-four feet long Canadian Rock-Elm piles that brought elm disease to Britain to concrete filled steel tubing, steel shuttering for sea defence, railway rails, reinforced concrete, and even a dear little bunker drill that came on the back of a small pick-up. It was erected in minutes by two men. They sat down and watched it all day, with its three legs and tiny petrol engine plopping away, fascinated.

Different piles need different treatment. You can hit steel a lot harder than wood, which will split despite the keeper ring at the top if you are not careful. Shuttering can buckle or bend over at the top, which is untidy if you want a neat edge for your harbour wall, although I have seen vibrating ones for shuttering. I think it must be difficult to get round ones in dead straight, judging from one major seaport I will not name, but I expect some of you can guess which.

Reinforced concrete is perhaps the most difficult if things don't go to plan. In hard ground the temptation is to hit them a bit harder, but since they don't move so far with each hit, the loading goes up and they spall; that is, scads of concrete blow off the sides just below the top, or anywhere down the pile if you are unlucky, or the casting firm has cut corners. Also, hard ground will give rise to a reflected shock wave, which arrives back at the top just after the weight has left it. With nowhere to dump its energy it kicks off the top few inches of the pile into thin air. The top then crushes with the next hit and you either have to cut it off with a big disc or thermal lance, or putty it up and hope the architect won't notice. What you really need to do is to hit it less hard, but more often, to keep the work rate up.

So how can we increase the rate for a piledriver that is powered only by gravity? Quite easily, it happens. What we need to do is to have two or more masses as the driver, instead of one, separated by a big spring. One could use a big steel coil spring or a piston in a cylinder with a pre-load of compressed air, which has the benefit of easy adjustment, or anything which is springy. In the days of waterpower they used rammed oak leaves to absorb the shock under trip hammers.

What happens then is that as the now divided mass comes down, the first mass delivers its blow, shortly followed by the second and subsequent masses if you have more than two. You get two or more hits for the price of one. As an added bonus, there is still a mass on top of the pile for the return wave, which delivers its energy and kicks that up into the air instead of the top of the pile. It actually delivers another hit down the pile, as it is reflected back down from the top.

Thus we can double the rate of working for a gravity driver by using two weights, or treble it for three, and so on, while at the same time moderating the blow delivered to keep within safe working limits, and keeping a 'lid' on top to absorb the reflected wave. And all for the cost of splitting the weight and inserting a spring.

The concept is not limited to piledrivers, of course. Upholsterers used to have a variety of hammers instead of the ubiquitous staple gun that they use now. However, restorers still use tack hammers; I have even met one who still spits tacks onto his magnetic hammer. Sometimes you can't just belt the tacks in with one blow, particularly brass-headed studs used for decorative work, so a split headed hammer would be useful, delivering two or more gentle blows per stroke, with less risk of deforming the tack head or bending the shank. Panel beaters could benefit from the device too. There are probably lots more uses that other people will think of as they go about their work.

We all like banging things in: clackety clack! Who wants a better piledriver, or tack hammer? All I want is one half of one per cent. Roll up! Roll up! A free go if you can get the bell to ring!

AN IMPROVED TOOTHBRUSH

This article is presented as a guide to the form you are likely to use in a patent application. At time of writing I do not have diagrams to hand however.

The present invention relates to an improvement in both the form and function of the toothbrush, in this instance specifically the interspace brush, although many of the improvements could be incorporated in other types of brush, whether used on teeth or not.

The novel features of the invention are that the brush has two heads, which are different from each other, as their function is different; one for the interspace and one for the gum/tooth space. The cross section of the bristles is different, so as to get different stiffness in different planes: the cross section of the bundle of bristles forming the heads, the surface features of the bristles employed being configured to improve their efficiency at removing plaque, and the cross section of the grip, being formed so as to determine the orientation in which the brush is used.

The function of the brush is to clean the space between the teeth, and between the teeth and gum. Such a brush is commonly called an 'interspace' brush, meaning that it cleans the spaces between the teeth, or more accurately the surfaces of the teeth adjacent to these difficult to get at crevices or spaces.

There is therefore a requirement for two different jobs to be done.

1) To run a brush along the crevice formed between the gum and the teeth, and to clean out the pocket formed there.

2) To clean that space between the teeth.

The gum pocket is thin and flat, therefore the head designed to do this job must be thin and flat and pointed, the flattening being at right angles to the body of the brush, so that it easily fits into the space. Also, as one is reaching behind the teeth, particularly the back teeth, an angle of slightly less than 90 degrees is preferable, and 80 degrees has been found comfortable to use.

The other end of the brush is designed to clean the vertical crevices between the teeth. These are of a different shape, and have a tooth on either side rather than gum on one side and tooth on the other. The brush head for this job is nearer to the normal round bundle of bristles, although there is of course the option of flattening it at right angles to the other head. As this is used with a different orientation, defined by the grip part of the brush, which will be described later, the preferred motion is a gentle massaging action, or a series of small movements of the head from the base of the tooth outwards. The best angle for this has been found to be about 100 degrees. These two angles are not highly critical; anywhere within a few of them degrees is serviceable.

To ensure that the brush head is presented at the correct angle to these crevices, an ergonomic grip has been designed. This has the form of an equilateral triangle in cross section, but with the sides slightly dished inwards. This falls naturally to a three fingered grip, such as one might use for holding a pen. This gives fine control, and is perhaps associated with delicate movements such as writing, whereas a full fist grip is associated with a hammer or scrubbing brush. More importantly, the flattened head that is used for cleaning along the gum-line is aligned so that it stands up vertically to the ridge between the thumb and first finger, and is therefore correctly orientated for use. The other head, being at 180 degrees to the first, falls in line with the opposed surface of the triangle, not being in line with a ridge. As the crevices between the teeth are at right angles to the gum-line, this ensures that this brush head is also correctly aligned for use from the lingual or inside of the mouth. There is a third angle of presentation, and this is used from the outside of the mouth.

The grip is symmetrical, thus it may equally be used by both a right and left handed person, or may be held in either hand for those people who change hands for different sides of the mouth.

Brush bristles, including those used in toothbrushes were originally natural products such as pig bristles, and synthetic bristles are copies of these natural products. They are true cylinders in section as this is easy to make by extrusion. However, this is not the best form for its function.

First we must consider the 'spine' or stiffness of the bristle. Obviously for a true cylinder of homogenous material this is the same in all directions. However there is an advantage in having a bristle that has more stiffness in the direction of travel, as it is its stiffness that causes resistance to travel and thus affects its ability to do work. Thus a flattened bristle will be more effective that a round one, whilst using less material and being smaller along the axis of the crevice and thus able to enter the crevice more easily. Thus both the brush head is flattened, and also the bristles forming that head, both being flattened in the same plane; i.e. the plane of the crevice. The degree of flattening is optional, but about two to one for the bristle, and four to one for the head used in the gum pocket, with not more than two to one for the spaces between the teeth, appear serviceable. The bristles are of course arranged so the flattening of the bristle is aligned with the flattened form of the crevice to be cleaned

Brushing a glass plate, and observing from underneath, allows one to see the working action of a bristle, or collection of bristles. One can see the contact point of the bristle, and its deformation under load by viewing through the glass. One thing is immediately apparent: the tip does virtually all of the work, and the tip is trailing. This is not very efficient. A surface formed specifically for the job it has to do is much more effective. The best kind of surface configuration can be found in the common file, or seen on a pitch block used for lens grinding. Here the surface is formed into a series of elongated diamond shapes, having angles typically of 60 and 120 degrees. Thus as the bristle is drawn across the work surface, instead of merely one trailing edge, a series of leading and trailing edges is presented. Further, the channels between these ridges form a transport network to remove material dislodged by the action of the ridges. It should be noted that the bristle need not be particularly harsh or abrasive to perform this function; one is merely removing plaque, or the bacteria responsible for the early stages of plaque formation, which are much softer than the tooth itself. For a normal toothbrush, where toothpaste is used, this action is even more pronounced, the channels retaining the toothpaste, which is mobile along them, giving a continuous presentation of the paste along a sliding edge. This is much the same as the way a soft pitch block presents jewellers' rouge to the glass surface of a lens, and removes the swarf produced in the operation.

Although the type of surface configuration described here is similar to that of a file, such surface features may be varied. For example, a series of parallel grooves could be used, and these could be formed by simple extrusion. Similarly, for the brush head used in contact with the gum on one side and the tooth on the other, the surface in contact with the tooth only could be formed, leaving the other smooth, for plaque does not stick to the gum, only the dentine or enamel of the tooth. For the bristles cleaning between the teeth, both sides would of course be formed.

The flattening, at a ratio of two to one or thereabouts, across the two axes of the bristle and the impressing of the pattern on the surface of the bristle could be done in one operation by passing the drawn filament through a set of heated rollers, a simple addition to the normal manufacturing process.

While these features are here specifically related to an interspace brush, they could also be applied, singly or severally, to an ordinary toothbrush, or any other type of brush.

For other brushes, such as paint brushes, a flattened bristle could be used, the flattening being in the same plane as the work surface. This would give a more even distribution for the paint and also provide a 'feed' as paint is removed from the tip of the brush; it would also be better for cutting in corners as the brush would not deform so much in the lateral plane. Longitudinal ridges could be used so that the brush will hold more paint securely, this speeding the work rate.

There are other brushes for various uses, which I will not detail here; they could all benefit from an improved bristle along the lines described.

ANTI - ANTIFOULING

If you visit the seaside in Britain, or many other countries in the world, you will find nearly every creek, harbour and inlet contains pleasure boats. These are in fact seldom used and are a considerable drain on the resources of those who own them, the pleasure being in owning a boat rather than in going out in it.

There is an environmental cost to this however, for most of these boats are hauled out and antifouled once a year. Thousands of tons of dangerous poisons such as mercury, arsenic, copper; and for commercial craft tributyl tin, leach into coastal waters every year with the consequent damage to marine life.

So how can we prevent this? Why, put the boat in a bag of course! A simple sheet of plastic, similar to a lorry tarpaulin with built in mesh ripstop, could be drawn underneath, with hooks to locate it to the rail, or a rope rigged in lieu of that A drawstring at either end to close the ends and pull the edges of the bag above the water, and the problem is contained.

A lot of these boats will already have their antifouling wearing off and will have marine growth such as barnacles, sea squirts and weed already growing on them.

Since we now have the boat contained, we might as well fill the bag with fresh water, pumping out the salt first. All the marine life will die off in a short while. This will then be much less well attached and could be removed by scrubbing with a high pressure jet. There will be no need to slip the boat; a pole with a jet, or perhaps a series of jets arranged like a rake, would do the trick. To prevent the reaction from the water jets blowing the rake away from the hull, plates either side could channel the jet, this will reduce the pressure by Bernoulli's effect and press it close to the hull.

We won't even need to slip the boat and spend hundreds of pounds on antifouling either. If we wish to use the boat all we have to do is to go sailing with our nice clean hull for a week or so, for growth will not matter much in that time, and then put it back in its bag. Some freshwater algae will grow, on the sunny side at least, but this is much more easily removed than barnacles, and will be smooth and offer little resistance in any case. The outside of the bag will still be prone to fouling, but if this is a problem we could reverse it occasionally. The water in the bag would become fouled with leachate, but this would slow as saturation was approached, and decline anyway as the antifouling was not replaced, while the small volume would make pumping out to a treatment plant an easy matter. A simple on-site treatment plant could be made from a filter-bed of iron filings, layered possibly in sailcloth sandwiches so the top could be replaced and the bottom removed as required. A small current could be passed to increase the electro potential. Alternatively, a crushed wood-charcoal filter could be used, again with a small current. This could be replaced, and the used section dried and burnt, preferably with some carbon monoxide in the tail-gas to keep it reducing, so as to deposit the metals. The exhaust could pass through a simple cooler to fractionate the copper, tin, mercury and arsenic etc. Tin is twenty odd thousand pounds a ton and copper a few thousand; mercury is not cheap either, so there could be a worthwhile return.

On a long cruise the local sailing club could hire out bags for people stopping for a while. Anodes used to protect the propeller and outlets will last a lot longer too, as there is very little electrolytic effect in fresh water, it being an insulator. For wooden boats, gribble, a boring mollusc, will be a thing of the past.

A mesh frame could be put over the engine-cooling inlet if the engine were run, to prevent sucking the plastic over it. Sink outlets are usually above waterline anyway so a line to draw the bag tight or a few inches of pushfit extension hose would suffice. You should not use the boat toilet in the marina, so that is not a problem. Rainwater from the deck could keep the bag topped up if there were spillage in or out due to small waves or wash in the marina.

The cost of the bag will be much less than a slip out, scrub and antifouling, and it will last many years. Perhaps we will see a crop of mussels growing on the piles up our creeks once more!

Antifouling manufacturers would hate you of course, don't try to license to them. The system could be franchised however, perhaps marina by marina, or by area, giving a rapid uptake; nothing much would need to be specially manufactured, because most bits are 'off the shelf'.

So here we have an established need and market, a low cost, low development time, fast to market invention, with great savings for the purchaser and extremely green credentials.

A PROPOSAL TO INCREASE PROTECTION FOR THE LONE INVENTOR

This article is a modification of an article first published in the magazine of the Institute of Patentees and Inventors.

At the end of this article you will see: © A Thompson 08. This shows that it is the intellectual property of the writer and the date of publication. Usually, in the case of a book, your publisher will lodge a couple of copies with the British Library and Library of Congress, and ascribe an ISBN no. to back this up. However, technically, the act of publication is disclosing to a third party, which can be a friend, as long as you can prove it later, that is! Another advantage of copyright is that it forms part of the author's estate and lasts for fifty years after death.

Design copyright is somewhat similar, but relating to the form rather than function; it may be a particular design for lace, an oft quoted example, or the angles and bends in an exhaust pipe, as a chap I knew found out when making cheap spare parts. In that case it was held to be a computer program for the item. Design copyright needs to be registered however, but is cheaper than a patent.

There is a sharp contrast between the protection for written work and invention, which works to the detriment of the inventor. While the writer establishes protection by publication, the inventor loses by doing so, unless submitted in confidence. For the inventor there is considerable cost in lodging a full patent (a provisional one is free for the first year), and this merely gives him the right to sue, which can be made very expensive by a corporation intent on theft. This need for secrecy prevents open dissemination and is a great hindrance to trade. Quite often a large corporation is not interested in an invention which will enable something to be produced faster, better or cheaper, as they have considerable capital tied up in equipment that will be rendered obsolete.

Quite often, the small inventor deals with minor inventions, and the idea of a 'petty patent' has been mooted, but with no result.

To explore the situation further, I wish to explore in more detail the protection given to the written word. If you say, "Think of a book," most people will think of a number of printed pages sandwiched between covers. This however is not the book; it is the way that it is presented and stored. The book is something less material, being a set of ideas, a coherent story, maybe technical information or an artistic work. If translated into another language, with a completely different set of characters, as in, say, Japanese, or stored in noughts and ones on a computer, it is still the same book.

My thesis is that a description of an invention, with any accompanying diagrams, is still intellectual property. There is no good reason for this to be lost merely by the act of publication, provided that the author claims intellectual property rights at the time of publication, and does not make it a 'public disclosure', a device sometimes used to prevent patenting.

To test this idea, I will now, 'by way of example' as they say, propose an invention, with the intention not to disclose it for all and sundry to exploit, but for interested parties to come to an agreement for its use. To treat, by way of an agreement, rather than to trick by trying to exploit for individual gain, since it is coming up to Halloween at the time of writing. The same applies to all the other inventions in the book; I still have intellectual property rights, which are in no way diminished by making the ideas available to all.

The invention is a simple modification of street lamps:

IMPROVEMENTS TO STREET LAMP REFLECTORS

The problem to be solved is one of wasted light and energy. Many reader will have flown at night; many cities re instantly recognisable from the air: London, with is vast sprawl of lights, usually dimmed by drizzle; Los Angeles, with its sharp terminator against the Pacific, or Nairobi, a bright jewel in the darkness of the bush. All very poetic, but what are we actually seeing? Mostly streetlights, and, as the name implies, they are meant to illuminate streets, not act as aircraft beacons. Pilots have satnav and radio beacons for that. It is immediately obvious that a proportion of the light escapes upwards and is completely wasted.

Further, if you drive down a well-lit road, you will be aware that the intensity varies; also, you may find an annoying glare from lights as you drive towards them, and may have to put the visor down.

To obviate this waste and reduce glare I propose that a reflector or refractor be fitted, so as to direct the light onto the road and spread out the light more evenly. The optics are simple; light spreads out as to the inverse square; so the reflector should be curved in such a way as to direct more of the light to the distant areas than directly beneath the light, and a cut off established so that it does not cause glare to oncoming drivers. Perforations could be incorporated to illuminate the area under the light so as not to put too much light there at the cost of less where it is needed.

One objection might be that the invention is obvious. However, obvious or not, it is not being done; there is a problem that is not being addressed. There would be a great deal of detail work; the design of different reflectors according to the height of lights, spacing and so on, but that is just detail.

So here we have a simple invention. I claim copyright of the form of words and any later application of the words in any material device.

So how would I get paid? By simple agreement, which need not be exclusive or expensive. As a guide, I suggest one tenth of a per cent as a guideline, without giving hostage to any other inventions, mine or other people's. Such a payment is not onerous, being a thousand pounds per million of sales.

In case you had not notice, such a suggestion is itself an invention. The concept of not giving up rights to all and sundry is itself an innovation. A method and a mechanism, if you like.

There would still be a need for patent agents to fine tune any agreements, the more so, as the proposals would open the floodgates for many inventions, cut red tape and speed up progress.

It is possible that some manufacturers might still proceed along the old line of taking without paying. However, if the intention is clearly set out at the beginning, and it can be shown that someone is profiting from the intellectual property without agreement, seeking their own advantage against that of the inventor, a clear case could be made.

It could be argued that an author's copyright is really for a story, a form of words, rather than any benefit that may be obtained from the reader, other than amusement. However, railway timetables are copyright and nobody reads one from end to end; they are for the extraction of information that is of clear benefit to the user, and there are other examples. Similarly, if one publishes a way of doing things which is novel, then protection should be given to the information it contains, and its use, not merely to the words only, as in a story.

© A Thompson '08.

I put this in, in case you need to licence an invention; you can adjust it according to your own needs:

LICENCE AGREEMENT

This licensing agreement is made the .... Day of.............20.....
Between......................................(hereinafter the Author) of...
....
....
.............and..............................Of.................
....
....
....

WHEREAS the Author is the author and sole owner of the intellectual property rights, whether they be patent, design copyright or copyright, subsisting in submitted, Vis...
....
....

AND WHEREAS the Licensee is desirous of taking a licence under the rights subsisting in the work;

NOW IT IS AGREED AS FOLLOWS;

1) The Author hereby confers on the Licensee the Exclusive rights in the United Kingdom or elsewhere to manufacture industrially any article design or device corresponding to the Work and to do all other acts to which a rights owner is entitled to do, including the right to grant sub licences for a term not exceeding the term of this Agreement.

2) The rights hereby conferred shall not be assignable by either party without the written consent of the other party.

3) The term of this agreement shall be for a term of ....years from the date hereof unless extended by agreement between the parties, save that the Author may determine this Agreement forthwith at any time whatsoever if the Licensee is wound up, fails to take any action re manufacture or sales, or if the Licensees business is transferred or sold or if the Licensee is more than three months in arrears in payments as hereinafter specified.

4) Where copyright is applicable, the Licensee shall mark, or require to be marked, each said article with the copyright designation (a 'C' inside a circle) followed by the Author's or any subsequent proprietor's name and the year of first publication.

5) All rights in the original and any derived work shall belong to the Licensor. The original material to be returned to the Author after a reasonable elapsed time sufficient for the preparation of plates or models etc. The Licensee to be diligent in their care whilst in their possession.

6) Prior to distribution and sale of any article using the work, the Licensee shall provide 6 copies for approval and also such copies as are necessary for lodgement to comply with foreign legislation regarding protection.

7) The Licensee shall maintain full and accurate records of all articles manufactured in relation to the work and shall permit the Licensor to inspect them at any reasonable time.

8) In consideration of the licence granted hereunder the Licensee shall pay a royalty of ....% of retail price per article whether or not

that article is sold, subject to a minimum of £...... per annum, and also a sum of £...... being an advance on royalty. Payment to be in £ sterling, or any other agreed currency.

9) On termination of this licence the Licensee will forthwith deliver a statement of all licensed goods in its possession and a statement showing all goods manufactured and/or sold immediately prior to termination on which a royalty has not been paid together with a cheque for the royalty then payable. The licensee thereafter having a period of 3 months to dispose of stock, at the end of which period the remaining stock shall be delivered to the Licensor.

10) Following the termination all artwork plates, prints or photographic or other masters of whatever nature embodying the work shall be forthwith delivered up to the Licensor.

11) The licensee shall render accounts to the Author at six monthly intervals, commencing on the first accountancy date after signing this agreement. All other royalties payable to be paid within three months of the accountancy date after which they will be deemed in arrears.

12) The Licensee shall be responsible for notifying the Licensor in any breach of rights occurs, and shall be diligent in defending the rights.

13) This agreement to be construed as being under English Law.

14) In the event of a foreign court decision, or a foreign government agency ruling affecting the contract to such a degree as to materially alter its intention, the Licensor reserves the right to revoke and re-negotiate the contract.

15) The Licensor makes no warranty that the patent or other rights are valid Worldwide, or that the manufacture or sale of goods hereunder will not infringe the rights of third parties in the UK or abroad.

16) The Licensee undertakes to keep all information, drawings and other material received from the Licensor under this agreement confidential during its term and for five years after its end.

17) Communications required to be given hereunder shall be deemed to be duly given three days after posting to the addresses of the parties set out above by prepaid registered post.

18) The Author reserves the right to enter the work for competitions and awards.

IN WITNESS WHEREOF the Licensor and Licensee have executed this document the day and the year first above written.

Signature of licensee ....

Signature of licensor ....

Witnessed by ....

Date...

SECTION 2

ALIGNED REACTIONS

Again in this section I expound some more inventions of a more general nature that could be of interest to people in those fields; they are a bit more difficult to develop, but quite often the difficult ones yield the best results.

The other day I was idly thinking about chemical reactions, as one does, and their similarities and differences when compared with atomic reactions. We have all seen those photographic plates with thin lines traced by the ionised reaction products of atomic fission, and I wondered if it would be possible to get the same sort of picture from the break-up of an energetic molecule, nitrogen tri-iodide say. Then I got to thinking about energetic reactions that are the result of combination of atoms, hydrogen and oxygen for example.

This immediately presented a problem, for whereas the break-up of a molecule is very Newtonian, with the different masses going off in different directions at different speeds, giving a rise in temperature as they clatter about and bump into things, you don't get this where two atoms combine. Yet the reaction products are observably hot. Obviously if water molecules are going faster than the original hydrogen and oxygen, there must have been a reaction mass to give them a boost, unless all the energy is locked up in vibration. This seems a bit unlikely, as they would merely vibrate apart again unless they shed some energy during combination. There is vibration, but this is usually a fraction of the energy, the rest being in translation and rotation, which sort out their own distribution.

The situation can be compared to having a 'pseudo-molecule', say two masses with a latch-and-spring mechanism on a low friction surface. Trip the latch and they fly apart, the pseudo chemical energy having been converted into kinetic energy. If you now have two magnets on the same surface, they will slide together, but on meeting they will be stationary in relation to the surface, although the energy released could be the same in both cases.

The solution I think lies in photons. Quite often in an atomic disintegration photons are released, and although massless, they do provide a reaction kick in proportion to their energy; not very Newtonian, but it happens. Light does exert a pressure. Chemical reactions that release energy also release photons, so we have a model.

Now, in examining charged particle tracks it is immediately obvious that they are directional, for fission into two particles can only give two directions of travel. Further, since they are charged they can be constrained by a field, so a beam can be produced.

Many molecules are polar; that is, they have a charge that is more obvious at the ends, like soap molecules or water. Thus, although in a chemical reaction normally all the molecules are in a random arrangement, I wondered if it were possible to line them up so they were ordered. One could for example get all the breakdown products going in one direction and photons going in the other, to produce an intense beam of light; since equal molecules will give equal photons, we have a chemically powered laser. But I have heard the term, so maybe it has already been invented? Or is that just the power source, not the orientation?

This could give rise to a much more efficient rocket motor. Rocket exhausts are very hot, and this means low efficiency, probably well under 10% from a look at the exhaust colour, for the only useful effect is the velocity of the exhaust, not the velocity of the random motions within it. Ideally we need a rocket with a cold exhaust, all of the energy being used in moving apart the reaction masses.

There are other useful tricks we could try with aligned reactions. For example, if we had the right explosive molecule, say a silver styphnate, or azide or any explosive molecule we could tag a silver or other metal atom onto, we could line them all up like soldiers on a charged silicon wafer, equally spaced in a kind of crystal lattice. Low temperature explosives are better; when an azide breaks up, you get high pressure nitrogen and not a lot of heat. Think of a balloon bursting; it goes bang, but actually the temperature is lower afterwards than before. We could detonate them, maybe by a laser flash, and inject silver atoms into the wafer, all with the same energy and penetration. If we wanted, we could blow a grid of holes into the plate, choose different sizes and spacing by selection of the explosive if we wanted, and make very efficient semi-permeable membranes.

By doping the plate we could apply a charge and have a selective semi-permeable membrane that would greatly increase the efficiency of electrolysis, or effect separation of similar sized molecules with a different charge distribution. We could also induce chemical reaction at the plate and use it as a catalyst. For example, in a molecule, polar or otherwise, there is a spatial charge separation between the protons in the centre and the electrons in the probability fuzz of orbits on the outside. By applying a charge we can stretch the molecule so that the protons and neutrons are no longer centred. If we can increase the field enough, less well held electrons are removed and we have an ion that is more reactive. Even if we don't strip electrons, the molecule will still be more reactive, on account of the lesser amount of energy holding the stretched electron, the field of another molecule at close proximity being enough to tip the balance.

Thus we can invent a whole new field of 'aligned chemistry', where molecules point in a common direction and deliver their break-up products in common directions. We can deliver very precise amounts of kinetic energy to exact locations and do it by the billion lot in a lattice, greatly facilitating production of nano machines, for a semi-permeable membrane is an example of a nano sieve. We could even mimic ion channels and ion pumps as found in cell membranes by suitable application of a field.

There are lots more options, from more sensitive photographic plates to hotter cutting torches. Perhaps aligned molecules could be partly dissected or built by application of laser energy, as different bonds could be targeted according to their alignment or shadowing. For very low energy reactions we might produce a radio wave emitter, as the photon energy and wavelength would be low. By suitable configuration we could print miniature circuit boards with atoms, perhaps using a polymer as a substrate and injecting metal atoms onto chosen sites. Ideally we need a rechargeable reaction to use as a printing press; load up the plate with a charge, blow them off by a reaction or charge onto the substrate, reload, feed next sheet and so on. Caxton would have loved it! We could use multiple layers, blowing holes and connections through a 3D printed circuit as required.

Anyway, chemistry is never dull; there are all kinds of exciting possibilities out there, just waiting to be invented, with perhaps a little funding to get them off the ground.

SELF DESTRUCT WEAPONS:

A RESEARCH PROPOSAL

All products, from cars to coffee cups, have a useful lifespan. For cars it is about seven years, for plastic coffee cups it is one use only. Unfortunately, having served their use, they then go on to what may be called their useless lifespan, which may be longer. If you are unlucky, the wrecked car (complete with two and half thousand coffee cups) may be dumped outside your front door.

'Planned obsolescence' is often used by manufacturers to maintain continuity of supply. We have all heard of bearings with 'lifetime lubrication'; well, I suppose it would be! Not all is bad, however. After all, we might not want three year old biscuits, or to be driving around in twenty year old cars. Also, there is now a trend for either biodegradable products or recyclable ones, and this is growing.

There is however one industry, isolated from cost pressures and public scrutiny, that has failed to move with the times. This is the arms industry.

If for example, you could go into your local supermarket and ask them to remove everything from the shelves that had an expiry date within less than a year, there would be a flurry of activity, and not a lot left. If you could do the same for a national armoury, they would probably look at you and pick their teeth. Yet food is more important than armaments.

Worse, old, disseminated and out of control stock is not just litter, but extremely dangerous, entering a new and sinister phase of its existence. Arms are deployed within a political timeframe, but they far outlast this, acquiring a negative value as they are lost, abandoned, or sold to enter a grey area of supply where they can fuel terrorism or minor wars.

What is needed is an item by item reappraisal of current production to see how the life after deployment can be cut, to take them out of use rapidly after a fixed period. The watchword should be 'Failsafe'. There are many routes to this end; detonators could cease to function, casings could degrade, the act of arming could also initiate degradation of the explosive by releasing a catalyst, metals could re-crystallise and become brittle, and so on.

The technology is fascinating and well within the competence of any college laboratory. The financial rewards are great, as old stock will be removed from the sales scene after a given lifetime; legislation may coerce compliance once new standards are set, and the patents devolving would be valuable intellectual property. The political rewards are greater still, for the removal of the constant threat would allow more rapid recovery for countries coming out of civil war, the amount of land out of use would steadily decrease, and terrorism would reduce due to lack of supply.

For those requiring more detail, see 'The Futurist' Sept-Oct 1993. pages 24-28.

WASTE STREAM UTILISATION IN FARM PRACTICE

The world's largest industry is farming. Although diverse, the practice is in essence based on the conversion of solar energy, via a metabolic cycle, to food and other products. Despite great increases in efficiency via greater yield, greater cultivated area, greater mechanisation and fertiliser input, more than half of the mass harvested is not utilised.

With any crop some part of the plant, be it stems, roots or leaves, is discarded. The principal reason is that while plants produce sugar as an end product of photosynthesis, this is then converted by condensation, the removal of a water molecule and the joining of separate sugar molecules into a chain, via starch to cellulose, the primary skeletal material in annuals. Cellulose is indigestible to humans and many other animals, except for ruminants that use gut bacteria to break it down.

If a simple process could be found to break this cellulose waste stream into useable sugar or starch, productivity would be greatly enhanced and in some cases more than doubled. This might skew production to an excess of carbohydrate initially, but substitution of carbohydrate producing crops for those producing oils and protein would occur as sugar demand was met from the new process.

There are several leads to be followed; obviously bacteria and fungi already break down cellulose. Hydrochloric acid also breaks cellulose bonds, inserting a water molecule and unpicking the polymer chain. Perhaps surprisingly, hot concentrated sugar solution also dissolves cellulose.

There are several phases to the program:

1) Locate an organisation, probably a University, capable and willing to undertake initial research, funded by a grant.

2) Choose a crop or crops, probably grain as it is in high volume production with established marketing, transport and storage facilities. Sugar cane and beet could be tried, but there could be a conflict of interest; success could destroy the industry in its present form, as the crops would become less profitable as other sources kicked in.

3) Examine all possible routes to cellulose breakdown, including small-scale third world initiatives in addition to large-scale industrial processes. This might be a bioreactor as an intermediate stage converter for animal feed.

4) Appoint a project co-ordinator to pull in more organisations and funding sources.

5) Proceed to pilot plant evaluation.

6) Obtain a Licence of Right so as to encourage further independent funding for the University or other body concerned.

The money would be used to initiate the program, probably to appoint one or more full time research posts to address the initial chemical problem, essentially one of an effective route to depolymerisation of cellulose to starch and or sugar.

The success of the program would be measured in increased world food production, on the same land area, the same fertiliser input and the same labour force, or the reduction of inputs once targets were met.

FOSSIL FUEL FACTS

Everyone has heard of fossil fuels: coal, gas, and oil; how important they are to the world economy, what we will do when they run out, and whether they cause global warming.

It might be thought that there is little more to be said, but there are gaps in the debate, which I hope to fill in to some degree. To recapitulate briefly, although the current focus of attention is on gas and oil, the major reserves are in the form of coal; about eighty per cent in fact. The coal mining industry is in decline, not due to lack of availability, but due to cost of extraction. Don't sell the family coal mine just yet though; we may well go to a methanol economy for fuel cells, and this is easily produced from coal by blowing hot oxygen and steam through a red-hot coke bed. If you get the steam hot enough, perhaps via an atomic reactor, you don't need the oxygen. The same plant will handle carbon from organic sources when the mines are finally closed, so you will have a head start.

Extraction cost is largely due to mobility, or lack of it. Gas is extremely mobile. All you have to do is drill a hole into a reserve, connect up piping, and you are in business; there is very little to do in the way of pumping, refining, waste disposal and so on, at least compared with oil and coal. This mobility is also one reason for its relative scarcity; geologically fracture a gas field, and you lose your gas, fracture an oil field and some of the oil will probably stay there, fracture a coalfield and nobody will notice.

There are two other reserves, usually discounted, that I would like to mention. The first is methane hydrate; this is formed when methane is produced underwater from organic decay. If the water is a few hundred feet deep, and not many degrees above freezing, the gas does not bubble to the surface as from your average bog. Instead it forms a loose association with the water, a sort of whitish ice, which is trapped in layers in the mud and silt it was formed from. Although it could be argued that this is not really fossil fuel, but part of the in-cycle biomass, some of it may have been there for quite a long time. The amounts are uncertain, and harvesting is difficult. The real interest however is not in its value as a resource, but in the fact that if we get global warming, it is likely to bubble up spontaneously. This might be a cause for concern as it is about one hundred times as effective as carbon dioxide as a greenhouse gas, which will cause more global warming and could lead to a runaway effect. It does not persist in the atmosphere indefinitely, as it will oxidise to carbon dioxide and water over a few hundred years, but it could cause a boost of several degrees while it is doing so.

The next reserve is peat. Again, one can argue that it is not a fossil fuel, for the top layer may still be alive, and the layers underneath are the past few decades of dead leaves. Go deeper, and in some reserves you are going back tens of thousands of years, so there is a grey area of definition. As with methane hydrate, it is not rated much as a fuel reserve; it has less than half the energy of coal, is wet and often a long way from where it can be used. There is quite a lot of it however, perhaps six hundred cubic miles. Everybody has seen pictures of tropical rain forests, and has heard how burning them contributes to global warming. In a rain forest, what you see is what you get. Nearly all of the carbon is in-cycle with very little leaf litter or humus, the temperature being high enough for bacteria and fungi to keep up with litter production. In a peat bog there is not much to see on the surface, it is all underground. There can be a lot more carbon in an acre of peat bog than in an acre of forest. Unfortunately, as it is drained and as the climate warms up, it oxidises, and may contribute more carbon dioxide than fossil fuel; much of a big swathe of land right across the Arctic, from Canada through Russia to China is tundra.

Thus these two under-rated resources may be traps for the unwary, making a significant contribution to global warming for little if any gain. To factor them into the energy equation one should really count the potential damage done by burning fossil fuel twice over; once for the fuel and once again for the unwanted side effects of peat and methane hydrate destruction. With these two minor items out of the way I wish to examine in more detail the process of formation for the major resources. Coal is generally reckoned to be formed from fossilised vegetable remains. Often leaf imprints and bits of tree trunk are found in, it so the evidence is not really disputable. Oil is more likely formed from marine deposits, and is derived from smaller algae; there are no major fossils, but some chemical signatures that point to biological origin. Ignoring the minutiae, both are the result of photosynthesis; that is, the splitting of atmospheric carbon dioxide to utilise the carbon for cell formation for plants, and rejecting the oxygen as an unwanted by-product. For the early plants it was actually quite useful, for it violently poisoned off the competition in the immediate vicinity. Quite a good idea if you live in a soup of competing bacteria.

Oxygen is a highly reactive gas. If you want to look for life as we know it on some distant planet, don't bother with traces of water, or carbon compounds. Look for oxygen. For extinct life, look for surface oxidation, principally of iron. For a long time, several hundred million years, oxygen did not accumulate in the atmosphere; instead the earth rusted. Luckily there was plenty of carbon dioxide, and eventually enough oxygen was left free for an ozone layer to form, blocking lethal ultra violet rays from the sun and allowing plants to get out of the sea and onto the land. The process of rusting has not finished however. At peak, in the late carboniferous, there may have been twice as much oxygen in the atmosphere as there is now. Some early dragonflies had a wingspan of three feet; if you bred them that big now they probably would not fly, as they would have difficulty getting enough oxygen to do so. Half of this oxygen has gone now and there is not enough carbon dioxide left to replace it. True, we are making lots, but that does not help, as we use atmospheric oxygen to do so, apart from a little from iron and aluminium production, where the oxygen comes from the metal oxide.

This leads to the obvious question: how much fossil fuel is there? The answer, surprisingly, is rather a lot more than we have been led to believe. Dire warnings of 'running out' of fossil fuel are way off the mark.

The trick is not to look for known fuel reserves, which tend to alter with time, but to look for oxygen. For every molecule of oxygen in the atmosphere there must have been one molecule of carbon dioxide split. CO_2 becomes O_2 in the air and one atom of carbon underground somewhere. We can roughly calculate how much there is. CO_2 has a molecular weight of 44; that is, it weighs 44 times as much as a hydrogen atom, which is taken as unity, being the lightest atom. Of this the carbon weighs 12 and the two oxygens 16 each. Very roughly we can say that two sevenths of the weight of the original CO_2 is now underground as carbon. We can very easily weigh the atmosphere; try to suck mercury up a tube by connecting it to a vacuum pump and you will get a height of about 30 inches, which corresponds to a pressure of about 14.7 lbs per square inch. One fifth or about three pounds of this is oxygen, which gives rather more than a pound of carbon per square inch of the earth's surface, seas included. This may not sound much, but it is well over a hundredweight per square foot, or over a million tons per square mile, and there are quite a lot of square miles to go.

We don't really need to go much further, for obviously if we burned all of this, there would be no oxygen left to breathe; in fact one per cent of carbon dioxide is lethal if you breathe it for long. Current levels are about four parts per thousand, but they were three less than a century ago and they could double in the next few decades. It could be argued that we might still run out because we could not find much of this hidden reserve, or the cost of extraction would be too high. There is a bit more left in the calculation however; to start with it is not only current levels of atmospheric oxygen we need to consider. There is as much again dissolved in the sea for example, and maybe a hundred times as much in those few million years' worth of rust.

There might even be some 'native' carbon, that is, carbon that never entered the biological cycle. Given this superabundance we are certainly likely to find enough; if we could bring it all to the surface we would be five feet deep in the stuff. The trouble is that we are likely to find more than enough, without looking very hard, to greatly change the atmosphere. So we are likely to have global warming. Once started it is likely to motor on a bit under its own steam even if we do try to cut back on fossil burning. Given that this is probably a fact of life, we have to ask whether this is such a big deal anyway; after all, perhaps the carboniferous might have been a great time to live, no ice caps, and warm seas! We could always shift a bit further north if the tropics became too warm.

Things might not be that simple however. To start with, we could not eat most of the plants that were around in the carboniferous, and our current crops don't seem to do too well under elevated temperature and carbon dioxide conditions. There would be a need for some pretty fast plant breeding; maybe not a problem, but remember that more than half of what we eat is grass seed. We are stuck with three varieties of one species for most of our food, plus rye, oats and barley, which do not add up to much compared with wheat, rice and maize. Worse, the sun is about a third hotter than it was in the carboniferous, so we can't go back there anyway; we would be more likely to boil the seas than have a steamy swamp. Also, if we abandon ten degrees, from latitude twenty to latitude thirty, the bit we might gain from latitude fifty to sixty is rather a lot smaller, as a quick glance at a globe will confirm.

So there might be a problem. The fossil fuel industry has got its calculations massively wrong; they are out by several orders of magnitude. It happens sometimes. Once we thought that the world was flat, and that stones could not fall out of the sky. The fact is that we shall run out of atmosphere long before we run out of fossil fuel, and there never was, is or can be a shortage. Since we cannot burn it all, we need to ask how much can we burn, and this may be close to what we have burnt already. Eventually we must come to a zero burn economy and need to be halfway there well within the next few decades; the length of the tail does not matter very much. Extraction for plastics could go on for a while longer, as most of this is re-fossilised in landfill anyway.

The probability is that we will not achieve this; there are too many vested interests, too many developing countries wanting to come on stream, a lack of public perception and a lack of political will. Thus we will to go to plan B by default. In this scenario we are quite likely to see a fall in food production and population levels. Doubling of the population every thirty-five years probably was not such a good idea anyway. If it halves in the same time period we need only to go back to 1980 levels and nobody much complained then. After all, you cannot talk to all of them and most people will know a few hundred others, with a dozen or so friends, whether they live in a community of a thousand or a billion. We might lose a lot of our cities to rising sea levels; 80% of us live in risk areas, but modern cities are in a constant process of redevelopment. Come back to your home town in a hundred years and only a few landmarks will be left, so we can shift redevelopment to higher land. It is a question of rate rather than total change, and sea levels are not rising all that quickly. We don't need a Canute mentality, or higher sea walls, we need a planned response to change. We can't save everything, so we must get good value judgements on what is more important. While the change might not be great fun it need not be a total disaster either. What is important is to save centres of knowledge and learning, for that is the core of a civilisation, not its numbers or buildings.

Plan B does not solve the problem however, it merely delays it and may make it worse. Halving the population will not make a very big impact on fuel use. Most of the missing will be in poorer countries that don't burn much fuel anyway, and those that are left will be richer. Moving cities might accelerate fuel use for concrete manufacture and building costs, so we still have to approach zero fossil fuel burning.

Luckily there are plenty of alternatives; we live near a star that delivers about seventeen thousand-horse power continuously per acre of land surface, and we know how to harness this. There is quite a lot of power available from wind and tides and we know how to harness this also. Biomass, principally wood, has always been used, although now not on a sustainable basis, and we can convert this to motor fuel, carbon or gas. We might even use nuclear power if we can achieve hydrogen fusion. More simply, there is a route to splitting water at only about dull red heat, using magnesium and iodine as a catalyst; much cheaper than hydrogen fusion, and just as useful. Other nuclear routes seem a bit expensive if you count the cost of tidying up afterwards and the unquantified risk of accidents.

The main argument against the use of alternatives is cost. This is a false argument, for the cost of the damage done by burning fossil fuel is never entered into the calculation. If we said petrol should be free, but one had to pay fifty dollars a gallon for the oxygen to burn it with and the damage done, the balance is greatly in favour of alternatives. I am not saying the cost of damage is really fifty dollars a gallon, but since both sides of the argument are unquantified it is as least as accurate as the current lack of calculation, and therefore has equal status. It could of course be a lot more if we lose the civilisation.

Another problem is that the new technology is emergent; that is, it is not fully developed. Comparative costs are not really known and tend to leapfrog downwards, they are not readily available, and there is great public ignorance. Many people, for instance, are still using tungsten filament light bulbs instead of low energy bulbs, which give four times as much light for the same electricity and give a return on capital of over a hundred per cent, while their savings languish at two per cent. There are many more examples. Low energy bulbs will go obsolete in their turn, and they do have a disposal problem.

Unfortunately, the largest industrial power base, America, is hog-tied by lack of incentives. What is needed is to get fuel tax in America rapidly in line with Europe so there is a level playing field; use the revenue as investment in new technology, and then we might see some action. Current arguments are that such a fuel price hike will damage American industry. Not doing so is likely to kill off large sectors of it rather than damage them. If the rest of the world develops alternatives and America does not, industry there is likely to be on a par with Russia at the end of the Cold War. Already there is a big gap opening up with regard to steel production, motor manufacture and agriculture; all of these industries are lagging in efficiency now. While they may be propped with subsidies for a time, eventually the day of reckoning must come.

Obviously we need a transition phase, and there are some good indicators here. For example, the current record for miles per gallon for a 'car', being a three or more wheeled vehicle capable of carrying one person at an average of fifteen miles per hour, is over four thousand miles. While not suggesting that we should all drive such spindly vehicles, we could get within two and a half per cent of this figure and drive a hundred miles on our gallon in the same time. All this could be achieved with bits off the shelf. Engine efficiency could be doubled with known technology, while weight could be halved relatively easily to give a further boost. Thus the first halving of fossil fuel use for transport could be achieved relatively painlessly.

Moving to fuel cells for area power needs and transport could give a further halving, by which time we are well on the way. Biomass could be coming on stream shortly after. We have some luck here also; the major limit to plant growth is carbon dioxide availability, as long as they have plenty of light, warmth and water, so there might be a growth boost working in our favour. They might be the wrong kind of plants that get ahead so far as the agronomists are concerned, but at least they will sequester some carbon. Deforestation and desert growth might negate this, but that is something we could try to control. Deforestation could be greatly inhibited by getting the price up; currently wood is more or less given away by the producer countries, leading to a 'use once and throw away' mentality on the part of the users. If producing countries levied a few hundred dollars a ton export tax, they would derive good revenue while waste would be reduced.

We could greatly increase plankton growth in the sea by fertilisation with iron, but this is something of a wild card and potentially risky. It could be used as a kind of last resort however, and would buy some time while we tried to get other things right. We would have to find a way of harvesting the plankton, or other things like krill that live on it, for plankton is very small stuff and mostly oxidises long before it falls to the bottom of the sea, so there is little sequestration.

Another factor in our favour is that chalk, limestone and coral reefs will dissolve more readily, sequestering carbon dioxide as bicarbonate; it was all dissolved in earlier times and has been deposited by living things as the level of carbon dioxide dropped. The process is slow however, and is more of a long term buffering effect over tens of thousands of years rather than a quick fix. Loss of coral reefs also means less fish habitat and accelerated coastal erosion as well, which is the last thing we want with rising sea levels. Currently things are going the wrong way as carbon dioxide acidification and rising temperatures kill off the coral.

At all events we are starting into an era of rapid change. The correct response is neither doom-saying nor defence of the old order. What is needed is some planning for an orderly transition, and perhaps an element of ruthlessness in carrying out change. If a few sacred cows drop along the wayside in the process this is no great matter. As jobs go in the sunset industries, more will be created in the new. Many of the old ones are actually wealth destroying industries if the full balance sheet is examined; there is little merit in profit for the few at expense of cost for the many. We cannot go back, so we shall go forward, willingly or not. The secret is not to be pushed from behind, but to be drawn by some reasonably accurate and achievable vision of the future. It could be a good one too; pollution largely negated, wealth and leisure increased, and a slowly improving natural environment rather than a destabilised one.

Of course, at present we are not really headed in that direction, and later means lesser; fewer people, less wealth, greater disruption and a longer recovery period. There is a potential worst case scenario of a total civilisation collapse if we don't take the carrot. It has happened before, and our present civilisation is certainly not immune. We are looking at a window of opportunity at present, with many of the precursors to change in place but not yet activated. Realistically, although we have a knowledge base that could reduce risk and adapt to change, we do not have the mechanisms to implement it. The root cause has been called 'the tragedy of the commons'.

I should perhaps explain this in a little more detail. There are essentially two systems of ownership in practice; personal and communal. Originally, as man developed, there must have come a time when the idea of personal ownership evolved, possibly from the instinctive idea of 'my partner, my children, my cow, my share of food' as being distinct from the rest of the tribe. Later the idea of personal possessions, such as clothing and weapons, came about as these things were invented. Most of the world, however, was either not thought of in terms of ownership at all, excepting that parts of it would have been held communally, such as 'our cave, our hunting range'. With the increase of invention came an increase of things to be owned. Trade and specialised manufacture created different classes in society, some of whom became richer and more powerful than others.

Commonly held land persisted until quite late even in Western society, the enclosure acts finally removing most of it from common ownership in the UK in the last few hundred years. The problem with commonly held land, particularly for grazing, was that a person could gain an advantage by putting an extra animal out to graze, although by so doing they might actually decrease the total yield. In other words, although apparently held in common, it was really up for grabs. We have exactly the same situation with fisheries now. Any country can increase its fish catch by putting more boats out, even if the total stocks are declining. Eventually the last fish is gone and we get a dramatic collapse. As we approach this collapse phase several things happen. Vested interests try to obfuscate the real situation, producing disinformation to show that they are being unfairly treated, that other people are to blame, that stocks are not really declining, or that the causes of such a decline are not overfishing if a decline is accepted. In addition, pressure groups will try to gain subsidies to limit the cost impact on the consumer, much of which will usually be pocketed as the just proceeds of the sale of votes rather than fish. Although everybody knows what is happening, paralysis ensues and the destruction of an industry, way of life, and trillions of pounds in value, proceeds apace.

The same situation is obvious in deforestation, fossil fuel extraction, and soil destruction by over farming and climate change. There are many more examples to be found in virtually every resource use. In all cases we have an example of some profit for the few, balanced by a far greater loss for the many. It is not a zero sum game.

Currently, the received wisdom is that we could sort out a technical fix, at least theoretically, but are unable to do so because of the lack of administrative structures, a basis of common international law, a lack of widely dispersed knowledge of the situation and so on. This may be a false assumption. The malaise may go a lot deeper; we may need a totally new ethos, a new way of looking at the planet, a new way of sharing responsibility. This is almost certainly unachievable, in fact the 'let's fix it' mentality of reducing carbon emissions, more efficient use of fuels, switching to solar power and in-cycle biomass is very unlikely to happen fast enough to make much difference in the mid-term. The idea that we will be able to discover some new way of thinking of things, the concept that some kind of ideal solution is 'out there', waiting to be discovered, is laughable from our present situation. It may or may not evolve as a response to some deeply traumatic collapse of our present way of doing things. The idea that we will be able to slip seamlessly and painlessly into some kind of New World where everything is better is very unlikely.

It has been tried before though. In the latter part of the nineteenth century, a 'dancing wave' spread down the pacific coast of North America. The native peoples from many tribes started to gather and perform a particular dance ritual. The authorities feared some kind of rising, tried to ban it, and a few people got shot. It was no insurrection however, but the results of a vision by a medicine man. Essentially it was the last kick of a displaced people against western civilisation. The essential elements were the performing of a dance correctly, as a result of which the surface of the earth would roll up, as though it were a carpet, and underneath all things would be made new. The waters, wrecked by accelerated run off and erosion caused by logging would run clear again, the fish would return, the land would be clothed in trees, and the animals would return. The scars made by railways, dams and towns would be gone, and significantly, for it was a stone-age society, there would be no more metal.

It did not work, of course, so they were wrong in their method of seeking a cure. They were perhaps not so wrong in their assessment of what was happening however. They lived there, and the concept of someone else coming through, wrecking and stealing and then moving on, was completely alien to them; one just could not treat the land that way and escape retribution. Actually, you can, so long as there is somewhere else to move on to, which is at the heart of our present problem. Now there is nowhere to move on to.

Rate is all. The country that gets ahead in the new technology is likely to stay there for a good long while and pull the rest after it. Historically this has been the case and there is no reason to suppose that it will not happen again. Quite often the young want to destroy the old and rebuild the world anew. This time we are likely to do the first part of the job for them. Let us hope they rise to the opportunity.

HOT ENGINES

In the future we are all going to use electric hydrogen fuel celled vehicles, and the internal combustion engine will be as dead as the Dodo. Right? Actually, wrong! The internal combustion engine will stage a fightback, achieving close to 80% efficiency, which is on a par with the fuel cell.

In this series of papers I wish to examine all possible routes to achieving efficiency, and the practical measures which will have to be undertaken to implement this.

This will require some innovative development. Not all companies will be able to make the transition in mindset required; a lot of big names may go to the wall, while smaller, and more flexible innovative concerns will achieve dominance.

The problem with the automotive industry in particular is that it is stuck in the first half of the twentieth century. Miles per gallon have not increased significantly since then, and most developments have been in cosmetic improvements and customer appeal. Safety improvements have been led not by industry but by legislation and pressure groups. Depreciation, running and insurance costs are rising, not falling: a sign of an industry in stasis, not dynamic growth. In short, we have a very vulnerable industry, likely to be caught short by events and overtaken by technological change arising outside the core areas of production.

This of course presents a great opportunity. Where there has been little change over a period of half a century, there is leeway for a great deal of change to happen very quickly and very great rewards to be reaped by the progenitors of that change.

Lear, an American inventor, spent a lot of time on steam engines, but was unable to beat the inherent inefficiencies associated with a relatively low temperature working fluid, heat transfer from combustion products to steam, and the loss occasioned by the rejection of the latent heat of steam in the exhaust. At a time of very cheap fuel this did not matter much, but prices rose, and running costs overtook his improvements. The steam engine is one of the most flexible types of engine yet designed however, and we will incorporate some of its advantages in the new design.

Big plant used for electric power generation has addressed the problem by tacking on steam plant to run off the exhaust from gas turbines, achieving efficiencies of 53%, a figure incidentally achieved by Vospers (a British shipbuilder) in the nineteen thirties by tacking steam plant for secondary machinery, mainly electrical generation, pumps and so on, onto the diesel exhaust on big private yachts for rich customers. Diesels were about 40% efficient then as now, while the steam plant took about 20% of the 60% left as waste heat. Not much new there then!

As a first example, I will re-examine combined plant. To do this we will start with a normally aspirated four stroke diesel engine, and then look at a series of incremental changes that we could make en route to higher efficiency. This might perhaps require teamwork, with different people working in concert at different parts of the problem, and bring in a wealth of technology, not primarily developed with engines in mind, to bear on the problem.

Our engine will probably have a thermal efficiency of around 40%: that is, 40% of the energy in the fuel will be used to produce power and the rest rejected as waste heat. The compression ratio will be around 22 to 1 and it will run on 'full air', control being effected by throttling the fuel, not the air supply. Metered air and fuel has already been dealt with in another paper, but the improvements suggested could also be applied to this engine.

Given that 60% of the energy of the fuel is lost via the exhaust pipe, the first thing we can do is to fit a heat exchanger, and heat the incoming air to the engine. This in itself will not increase the efficiency; in fact it looks initially like a thoroughly bad idea. To start with, the mass of air entering the engine will be reduced, leading to a decrease in power. Also the engine will run much hotter, as the increased temperature will lead to an increased exhaust temperature, which in turn will be recycled to an increased induction temperature. We are likely to get meltdown in a short period of time, so we have to do something else as well. Also, there are some control mechanisms we need to put in place. First of all, the density of the air, and hence the amount of oxygen will vary as to the induction temperature, thus we need an oxygen sensor in the exhaust to keep the burn near stochiometric, say 3 to 4% excess oxygen to prevent smoking. Next, we need to monitor the combustion temperature to prevent it going too high. We are also likely to get more oxides of nitrogen. These are slightly endothermic, robbing a little heat from the burn, but the real problem is that they are pollutants. We will have a catalytic converter in the exhaust however, which will return the heat again, so there is not much of a loss.

On the plus side, an engine running on preheated air will be able to burn all kinds of recalcitrant and difficult to burn fuels not normally considered as suitable for a diesel engine. Once the engine has warmed up, we use metered water or steam injection every alternate stroke; this has the effect of keeping the temperature within the design parameters for the engine, and also providing a free power stroke from what would otherwise be waste heat. Since we are combining a steam and diesel cycle in this way, we no longer need a water jacket and radiator. In fact we need to wrap the engine in insulation to prevent loss of heat, just as we wrap steam pipes in other plant; the water jacket would thus become a boiler. While we need to address the amount of fuel injected, to keep a stochiometric burn, we can inject as much or as little water or steam as we like, thus giving a very fine control over engine temperature. We could also vary the ratio of fuel to water injection strokes, possibly running one in three or some other figure if need be. We could also look at injecting water into the combustion stroke. This could lead to 'quench' and the passing of unburned fuel, so care is needed. It would be particularly suitable for an engine burning alcohol, as this is fully miscible with water. Gas turbines have been run on as little as 15% alcohol and 85% water. Not much power available though! If you want to explore this route, design a gas turbine that will burn nearly all the oxygen passing through it, with the lowest possible temperature rise, then use it to power a water spray, also injected into the spent gas, and use it as a fire extinguisher. A vast amount of vitiated air injected into the base of a big fire would be more effective than dumping high-pressure water on top of it, particularly for oil fires. Make sure you clear the area first however, and that your firemen have oxygen or compressed air breathing apparatus.

To return to our engine, we could inject the water a split second later into the hot combustion gases, a kind of stratified charge. This would have the added effect of increasing the pressure on the flame front as it burnt down the engine. We might have to look at fuel injection also. Injecting into very hot air might give a shaper burn, with the potential for knock or detonation, so a slower injection might be needed.

Very hot engines don't run well. There is a problem with lubrication, excessive wear, piston meltdown and distortion. We could try directing an annular water spray onto the cylinder walls as well as through the hot gas; we might even use a sintered, micro porous wall with a water jacket outside it, steam bleeding through as the pressure rose in the waterjacket. Steam and hot iron are uncomfortable bedfellows however; we are likely to get iron oxide and hydrogen if we are not careful, so perhaps stainless steel would be in order, or better yet, a sintered ceramic. This has very low expansion and if prepared completely free from voids it can be very strong and unexpectedly flexible. Having solved the wall problem, we can now run red hot if we want to. But now we have a lubrication problem; oils tend to carbonise, break down and oxidise if they get too hot. So we will have to look at silicon oils, a whole new product range to be explored.

Thus we have a whole range of bits of new technology, to be addressed by a series of graduated steps, leading to a very much more efficient engine, with sensible gains in the early stages. We will have effectively blended the advantages of steam and diesel technology into a seamless whole.

Next, we will address power-weight considerations. But that is another story!

BACK PRESSURE

Have you ever wondered what it would be like to be inside the crankcase of an engine and see the pistons going up and down from the other side?

Well, perhaps not; it does after all, require a little imagination. For the average car engine there would hardly be room, unless you imagined yourself to be the size of a beetle. For larger engines, such as are found on ships there would be plenty of room, enough to hold a party in fact. Big engines can be as tall as a three storey house, and the cylinders big enough to hold three men at once, chipping the coke ring off at top dead centre, or so I was told by someone who had performed that operation. He also said the piston rings were neoprene rubber, but maybe he meant there was a buffer on the lower one, although of course it would explain the coke ring.

Big triple expansion steam engines never had a crankcase. Once I went on the Waverley, a triple expansion steam paddler that tours the seaside holiday resorts round the UK. You could see the connecting rods pushing the crank round, and the eccentrics of the valve gear. They had a first, second and third engineer too, all in spotless white coats, with a chair each. I suppose they had a piston each. Every so often one would get up and put a dab of oil on some vital part in an offhand sort of manner as though they did that kind of thing all the time, which I suppose they did. Personally I would have fitted a screw down greaser with a big lump of suet in it and changed it one a week, but things have moved on. Probably you can't get suet so easily nowadays. We used to have suet pudding quite often when I was young. It strengthened the arteries.

One thing you would notice in your car sump is that just as much air is displaced underneath the piston as there is above it. Also, with piston speeds of several hundred feet per second, it would be moving quite fast, and this seems to be rather a waste of energy. Exactly how much energy could be ascertained by motoring an engine with an electric motor and checking the power required over a speed range, and then pumping out the sump to near vacuum and repeating the experiment. You could do the same again with the space above the pistons evacuated as well, to see what losses were mechanical, by way of sliding parts, oil pump etc., and which were air pumping losses, and so get a far better idea of where to look for savings. You could even test each bit of the engine, part by part, to find out where all the energy was wasted, instead of spending ten times as much money to make it more powerful to overcome the inefficiencies by brute force. Henry Ford did the same sort of thing with his engines to find out which parts never broke, and then made them less strong to save costs.

If we did try to run an engine with a vacuum sump we might run into other problems. There would be no mist lubrication for example, although we could go to direct spray lubrication, there being no air to break up a well-directed jet of oil. Also we would probably be surprised at how much blow by there was at the piston rings. The oil would vaporise at a slightly lower temperature also, and any dissolved gas would come out and maybe froth.

We could of course go in the other direction, and run the sump at a few hundred pounds per square inch. It might be best to use parallel motion, doing away with the small end and having a connecting rod solid with the base of the cylinder, sliding through a gas seal. Air would not be pumped turbulently from one cylinder to another but every piston would effectively have a return air spring under it. That way we could reduce the need for balancing weights, high inertia loading, and a host of other problems. The ideal would be to cycle as much pressure under the piston as on top, giving a very fast pick up for the engine and a great deal of weight saving. It would probably stop a lot of rattles and vibration too.

It is not such a big step from there to go double acting as well; after all the steam engineers did. Why waste all that space when everything else is in place and waiting for a doubling of power? If you were a really keen engineer you could take another route and use the exhaust from on top of one piston to power under another and have a double expansion engine to boot. When petrol is twenty pounds a gallon, you will wish you had done so, and had an engine all ready for market, years ahead of the competition.

A FIRE IN THE BOILER

Most people think of the age of steam as being the nineteenth century, however it really began in the seventeenth and is still with us now; our modern nuclear power stations are really modified furnaces still running the same old steam turbines. Even the latest gas turbine stations have steam turbines to extract the last available energy from the tail gas. Efficiencies are around 53%, a figure obtained in some private yachts in the 1930s by running steam plant off the diesel exhaust. They are in a sense cobbled together from existing technology and not designed from scratch. In the future, the concept of putting a steam boiler at the tail of a gas turbine may seem hilarious, rather the equivalent of using a mine pump to raise water to run a water mill, the first use of steam to drive machinery. However there were very good reasons for using steam in this manner; the water mill already existed as a complete functioning unit, merely disabled by drought, which could be easily rectified by pumping the tail pond back into the head pond; a practical rather than an engineered solution. The invention of the crank soon had the nodding beam engines driving belts to power looms, while steam pressure was used as well as atmospheric pressure, making the engine double acting. The beam gave way to the crosshead and the industrial revolution took off in well-ordered logical steps, with the inevitability of a Bach fugue.

Towards the end of the nineteenth century some minds were becoming dissatisfied with the low efficiency, weight and slow fire-up time of the steam engine, and another inventive step was taken. Why not throw away the furnace and boiler, and put the fire straight into the working cylinder?

A panoply of gas and oil engines was developed, with or without a compression stroke and having varying means of ignition. Most are forgotten now, but two contenders emerged; the relatively low compression spark-ignition petrol engine and the high pressure compression-ignition diesel engine. The diesel is gaining ground now due to its greater inherent efficiency. As compression ratios rise for the petrol engine, and injection becomes the norm rather than carburetion; the two forms may merge, the more so as the diesel becomes lighter and faster running. Elsewhere, I will show how this can be further increased from its current forty per cent to over sixty, but that is not the subject of this chapter.

What I wish to consider is a further, or perhaps sideways step. Having got the fire out of the furnace and into the cylinder, and found the drawbacks associated with this, I wish to put the fire inside the boiler!

The boiler of course will not contain water, merely the fuel and a forced supply of air; it will be a pressure cylinder for the combustion of difficult fuels, such as corn husks, chaff and straw, sawmill waste, grape pips, household rubbish and maybe intractable oils and tars if we can clean up the burn by including chalk dust to get out sulphur in one easy stage.

The immediate problem that springs to mind is that of the early steam engineers as they moved to high pressure steam. If they got things wrong they had a highly explosive item on their hands; the development of safety devices became a high priority. When I was young the Gosport to Portsmouth floating bridge was a source of wonder. A huge cogwheel with apple wood replaceable cogs that would have looked better in a watermill drove the winding gear that wound the chains as it clanked its way across the harbour mouth. The low-pressure 1831 boiler had a safety valve that was designed to let air in when it came off steam, to prevent atmospheric pressure crumpling the boiler like a tin can. As pressures increased safety features multiplied; the Stephenson governor prevented the engine running wild, while a weight operated safety valve was backed by a separate spring operated one, with a fusible plug ready to melt down if the temperature went too high. The boiler inspectorate gave the plant the equivalent of an MOT. Nowadays control is all electronic of course, but designers please note; the current supply should still be backed by a separate circuit to an emergency generator, itself backed again via a separate wiring loom to a battery. Fusible plugs and the like should still be left in place. They work when all else fails.

The containment of a high pressure, high temperature gas is not something to be taken lightly, so we too must put safety first. To begin with we will make the combustion chamber as small as possible, and then we must insulate the walls from the searing heat inside, for our steel will soften if it becomes red hot. Later we might go on to more exotic metals, but we start with what comes to hand; a modified gas cylinder such as is used for carbon dioxide or a scuba diver's cylinder might do. We could use normal fireclay to line the inside before going on to foam silica. Externally, we could cool the walls by air, or possibly water if we wish to inject some to drop the temperature by reacting the hot carbon inside to form hydrogen and carbon monoxide. A containment jacket could be used in case the worst happened, possibly blowing down to a 'quench' - a water trap that would extinguish the expanding fireball.

Fuel will have to be reduced to a useable form; sawdust will need little processing, while our brushwood, old cardboard and corn stalks will need chipping. A screw feed will drop this into a ready use hopper, through which will run a ram, lifting to allow fuel to drop into a small cylinder, from whence it is rammed into the combustion chamber via a slide valve with a chisel edge, in case any odd bits of tough wood or metal try to prevent it closing. A similar device can remove the ash at the lower end of the cylinder. For most plant material there will be little ash, about one hundredth to a thousandth of the volume put in for vegetable material. It is likely to be molten on extraction, the principle constituents being silica and magnesium and calcium oxide. We could employ additives to fine tune the end product. If we release the pressure explosively we are likely to get a fine broken rag of material, like flock. This can have a value as insulation. The material is in some ways equivalent to lava: a lot of dissolved gas bubbling out breaks up the material, a little leads to a more solid product which can be like toffee before it sets. Thus, by releasing gas slowly we can vary the characteristics of the product. Steelworks do the same sort of thing, although they have to add air to expand the slag, and handle much greater quantities due to impurities in the coal and iron ore.

The pressure inside the cylinder will force the valve against its seating, but you can have a two stage entry if you like, to prevent blow-out in event of valve failure. The fuel will drop into a fierce environment. The furnace will run far hotter than a normally aspirated one; we may even have pressures of two or three thousand pounds per square inch in later versions, but will start our experiments much lower.

We also have a point of divergence here: do we allow combustion to go to completion, or rather use our cylinder as a pressure cylinder to produce fuel for a more conventional engine? Both options should be followed; if we go for complete combustion we have in effect normal steam plant without the steam, the engine working off a very hot high pressure gas instead. Efficiencies will be much higher, in excess of a diesel, with the advantage of using fuel that might otherwise go to landfill. If we go to the other option (and we should consider oxygen injection here, to obviate the nitrogen in the air diluting our product,) we have the possibility of a small plant that operates continually at maximum efficiency, generating gas at night that can be used during the day, or piped to a variety of different engines, or sent down the gas main, or bottled. Again, we are taking in low quality fuel and turning it into a much more useful, easily transported product, the more so if we put in an electricity generating plant and feed the grid.

We will have to have a compressor to force air into our pressure cooker. It does not matter that this will use energy, for in effect the energy will still be there, and capable of being re-used by reason of the pressure and temperature generated. If we draw off gas from our cylinder for use in another engine, we can also draw off compressed air from the compressor, thus obviating the induction and compression stroke, giving a highly compact powerful engine yielding a horsepower for every couple of cc swept volume, although we may prefer to expand down to near atmospheric to extract the maximum energy from each cylinder full, trading power for efficiency. Ignition will hardly be a problem; too high a temperature might be, so we could inject water instead of fuel, or a mixture of the two to keep temperatures down. Efficiencies are again high. If we send the gas down a gas main we will lose some power as the pressure is dropped, but we could run this through an expansion engine without combustion to regain most of it and cool it at the same time, for a gas doing work is cooled by expansion, as opposed to a gas merely expanded without doing any work, where there is no temperature drop.

If we step back and look at our outline plant, the rough sketch of ideas before we get down to the nitty gritty, we will see that we have combined a slew of nineteenth century practices into something new for the twenty first. We still have pistons in cylinders in our engine; they have been there since it was a Roman pump, two millennia ago. Now however we have a modified steam engine, no longer using relatively low temperature steam, but burnt air like an internal combustion engine, although in fact we are an external combustion one again. We can still have combustion in the working cylinder if we wish, but have done away with the induction and compression stroke, simplifying the engine and more than trebling the power. Alternatively we can use our gas in a gas turbine. While we are at it we could look more closely at a steam turbine and see if we can modify that to run high pressure gas, with perhaps a power take off stage before combustion to prevent outrageously high pressures and temperatures. Perhaps we should go to the potteries to develop ceramic turbine blades capable of withstanding high temperature combined with a low coefficient of expansion.

Our new style pressure cooker is a modified Victorian gasworks, crossed with a Papin digester, capable of handling any fuel due to the much greater temperatures obtained. Steam injection can reduce solids to gaseous fuel that is easier to use and transport, or we can modify this simply by passing it through a catalyst bed to convert it to liquid fuel, principally ethanol, a useful liquid as a starting point for further chemical synthesis if we do not wish to burn it. We can of course take our key idea, that of high pressure combustion, off in other directions. For example, we have in essence a high pressure blast furnace. Blow in oxygen instead of air and use high quality anthracite and iron ore instead of grape pips, and we can produce steel. Unlike a normal steel works which consumes vast amounts of electrical power to run the blowers and the rollers, we will actually produce a power surplus which can be fed to the grid. While we are at it, we could look at the steel produced. It is likely to be full of bubbles. This is normally considered a disadvantage, although we can get them out. However, what if we leave them in? We might be able to get a new product; foamed steel, a lightweight rigid construction material that has considerable advantages over the old fashioned, heavy, solid, flexible stuff. Luckily, iron, on cooling, undergoes a phase shift. Cool down a piece of cast iron until it is dull red, and it is likely to suddenly glow bright red again as the crystal structure shifts from one type of arrangement to the other. Do your popcorn trick then, and there are a variety of possible products according to what amount of which phase you freeze in. We could also look at foaming other metals too; aluminium is a prime contender. We could of course try other gases. Iron has a particular affinity for carbon monoxide, although most others will dissolve also. We might not use dissolved gas, but inject gas into molten metal. Some liquids do not foam easily, or the bubbles break down quickly; a foaming additive might be needed, or injection into the metal when it is beginning to set, with a crystal lattice in place.

We actually have a big research program to define a table of pressures, temperatures, gases, and additives to aid foaming ahead of us. We have done it for plastics though, at much lower temperatures and pressures, so some of the groundwork has been done. The work will be costly, and take decades, but the rewards are huge.

Here we have the bones of invention, a series of graduated steps that are easily understandable for someone used to the old way of doing things, using off-the-shelf bits and pieces and off-the-shelf technology, but ending up somewhere new. Too radical a step may run into unexpected technical problems, while psychological resistance may be felt. One can imagine the old-guard resistance to the first clay water pot in the corner of the cave or hut.

'Why can't you go to the river like everybody else? And anyway it tastes of mud!'

The potter won out though, until the advent of water pipes, which of course leaked and got silted up. Or so the potter said. Nowadays we all drink bottled water, with magic properties, imbued by advertising, although most of it comes out of the ground in the same way as ordinary water. 'Eau de Robinet' sounds chic and impresses our friends with our taste and sophistication, or at least the non-French speaking ones.

Actually, although the scientist may have an image of himself as doing highly important work, adding to the field of human knowledge and increasing the quality of life for all, much of the effort is subverted to serve fashion.

High density information storage? Great! What information? Music mostly, and people stick little loudspeakers in their ears and go out to get run over in droves because they cannot hear the traffic, or video.

My aunt had a television set before the war. She told me the programs were much better then. On reflection, she may have had a point.

New car design? Actually not. Mostly new designs consist of new sheet metal, which is highly visible, while the engine does the same mpg as it did fifty years ago.

New crop species? There aren't any of note; we are still messing about with those our Neanderthal ancestors chose. We are actually losing species at an alarming rate, while growing the old stuff with decreasing efficiency and increasing subsidy. It is strange to reflect that food, which people will continue to buy long after they have given up on a new car or television has a subsidy. It would make more sense to subsidise things that are difficult to sell, if one has to have subsidies at all.

Thus although invention makes the society we live in, with its manifold advantages, society also feeds back into the invention process, and conditions what will be successful or not. We live in a flux of shifting perceptions and values and not all invention is good. Quite often there is a pressure to continue with more of the same, instead of a rethink about what we are doing in the first place.

A HIGH EFFICIENCY INTERNAL
COMBUSTION ENGINE

Fuel prices have risen sharply in the past few years, and environmental concerns have become more pressing. In addition to buying more expensive fuel, large consumers may also have to buy carbon credits also, thus making greater fuel efficiency of primary concern. 'Business as before' thinking is no longer a viable option, for we are as yet only at the lower end of an increasing price regime.

To achieve this, I propose to examine piston engines, petrol and diesel, both four-stroke in their simplest form, and steam, and to extract the best design features from each. The treatment will be non-mathematical, this best being reserved for a future stage when one is considering design parameters and cutting metal, for there are several interacting considerations which need to be brought into balance at this 'trade off' stage of design.

Petrol and diesel engines are in some ways similar, and current design trends such as fuel injection and higher compression ratios are bringing them closer together. Differences are more indicative than similarities however, and these are as follows.

The petrol engine is throttled: i.e. there is an air control as well as a fuel control. This is needed as petrol only burns efficiently within a certain air-fuel ratio; too weak a mixture and flame propagation will be too slow, leading to loss of power and hot running as the fuel may still be burning as it exits the cylinder. Too rich a mixture will also lead to loss of power and fuel waste due to inefficient burning, plus sooting up of the exhaust system, as the hydrogen tends to burn off first, leaving carbon. Carburetion, the provision of a correct petrol-air mix by means of a throttle cut-away and Bernoulli tube, has always been difficult to optimise over a wide throttle range, hence the trend to fuel injection, where it is sprayed directly into the cylinder, as in a diesel.

The disadvantages of a petrol engine are firstly an upper limit to the compression ratio, for at much above a ratio of 14 to 1 (special fuel excepted) the engine is likely to suffer from compression-ignition, which will lead to knocking and consequent loss of power. Also the throttling of the air supply results in a lower pressure in the cylinder and reduced burn efficiency. It should be pointed out that the compression ratio of an engine is that ratio between the swept volume, traced out by the piston, and the clearance volume at the top, not the actual compression ratio of the gas above atmospheric. If the cylinder contains air at below normal atmospheric pressure at the end of the induction stroke, then the pressure and temperature will be correspondingly less at the end of the compression stroke. Thus the advantage of having a pressure and temperature in excess of atmospheric at the start of the working stroke is diminished, for the efficiency of an engine is directly related to the initial temperature at the beginning of expansion, and the end temperature at the end of expansion, so it is important to match the engine to the expected load. This is why most large-engined cars return low m.p.g. figures. The large engine is not inherently less efficient, but it is just not working hard enough.

The thermal efficiency of a petrol engine is generally reckoned to be about 30%; i.e., 70% of the total energy of the fuel is rejected as waste heat.

The diesel engine is not throttled; it always runs on 'full air', control being gained by varying the amount of fuel injected into that air. It is not generally realised that the diesel is not an expansion engine, or at least there is no net expansion over the working cycle, for if one litre of air is taken in by the close of inlet, then one litre must be ejected at the close of exhaust, expansion then taking place in the silencer. There is of course compression followed by expansion in the working cycle, but these are equal, power being provided by the increase in pressure on the expansion stroke, due to the burning of fuel. For an air throttled engine, a lesser volume is taken in at induction as compared to that expelled at exhaust.

This is one disadvantage in regard to efficiency, but there are others: Due to the reduction of fuel only, the burn departs from stochiometric. A stochiometric burn is where there is exactly enough fuel to combine with the oxygen present in the air, for this will lead to the highest attainable temperature and pressure. In practice about five per cent more air is generally allowed, as combustion is unlikely to be complete in the short time the fuel-air mix is in the cylinder, due principally to variations in mixing.

The efficiency of a diesel is higher than for a petrol engine, despite all this, being generally regarded as being 40%. For convenience, you can use the figures of 15 H.P hours per gallon for spark ignition, and 20 for a diesel.

For both types of engines, more than half the power available in the expansion stroke is used up in the subsequent compression stroke. This seems to be unavoidable, for one must have compression, and also this amount of energy is returned in the subsequent expansion stroke. However it is perhaps worthy of examination to see if there are any ways to reduce this energy cost.

In a steam engine, from which both the above types were derived, the piston acts in the cylinder as before. However, as there is no need for a compression stroke, the pressure being provided by the boiler, the engine in its simplest form is a two-stroke engine. Also, there is little or no clearance volume, commonly called headspace; the piston travels to the end of the cylinder at the top of its stroke. In practice of course there is a little space due to the need for tolerances for expansion. A small amount of 'cushion steam' may be left in the cylinder to act as a buffer also. Steam is admitted to the cylinder, usually by a slide valve, and throttling is effected by variable cut-off for steam supply. The steam is then worked expansively, often through two or three cylinders of progressively larger diameter, before being blown down to below atmospheric by means of a condenser.

In many ways this is the ideal type of engine, very efficient over the working cycle, with an infinitely variable throttle, a high power weight ratio, with a great reserve of emergency power if worked non-expansively at the cost of efficiency.

The major drawback of course is the need for a furnace and boiler. Once this is added the weight advantage is lost, as is also overall efficiency due to the relatively low temperature of the steam compared to the combustion products in an internal combustion engine. Also, there is a loss to the condenser due to the high latent heat of steam.

The efficiency of a steam engine varies greatly as to the type of engine, but is seldom more than 20% except for large turbine plant running at pressures of over 3000psi.Steam is however making a minor comeback by way of scavenging heat from the tail gas from gas turbine plant. Here overall efficiencies of 55% may be obtained from the combined cycle. Some private diesel yachts of the nineteen thirties also used steam to scavenge heat from the diesel exhaust, with similar efficiencies. However, this dual plant, belt and braces technology is a bit expensive.

So how can we increase the efficiency of the internal combustion engine? The best way is to combine its best features with those of the steam engine. This can be achieved by running from a cylinder of compressed air, which substitutes for the boiler of a steam engine. This air could also provide for starting, horn and so on, thus obviating the need for a heavy battery, although we would need some power for initial warm up and lights etc. Provision of this compressed air can be done far more cheaply in energy terms than in the working cylinder of an internal combustion engine, for compressing air in a hot cylinder is not a good idea, nor putting cold air into that cylinder in the first place.

We compress the air isothermally, as nearly as possible; that is, without a temperature rise. This can be done by a multi-stage compressor, or perhaps by trickling water through the air as it is compressed. A sort of shower head at the top of the cylinder would do, the water carrying away the heat of compression. However, we are unlikely to get ignition if we use cold air in our cylinder. No problem! There is plenty of waste heat available from the exhaust, so we interpose a heat exchanger in line with the air entering the cylinder. We could even have a pulsed system, for the cold air from the compressed air cylinder will expand many times as it is heated, giving an energy increment.

Next, we could have linked metered air and fuel; thus maintaining a stochiometric burn at all throttle openings. We have an additional feature in that under heavy load the engine could run non-expansively, running down the air cylinder to provide extra power. About one horsepower per two c.c. could be achieved, although not for long, unless one had a rather large air cylinder. Under light load, the cylinder could be recharged, thus giving a more even running load for the engine, always a desirable feature.

The net result of all these improvements is that we would achieve an efficiency equivalent to a Sterling engine, without the need for external combustion and consequent slow running. Net, it works out at about 60% efficiency as compared to 40% for a diesel engine.

This would bring additional gains; for a cargo vessel, the reduction in bunkerage weight by a third would increase cargo capacity, or mean fewer refuelling stops. The extra cost of the engine would be repaid many times over its lifetime.

Historically, engines have started out as static plant, and then gone locomotive later. Probably the same pattern will apply, the first use being a stationary engine for electric power generation, where continuous operation would repay cost rapidly; later, trains and boats would come on stream, then heavy goods road transport, with cars after that. Later developments might embrace ceramic cylinders and pistons, with a high temperature silicon oil as lubricant.

A HIGH EXPANSION ENGINE

In a normal internal expansion engine, at exhaust opening there is still considerable pressure in the cylinder. This represents a waste of energy as the residual gas expands explosively, as can be heard if a silencer is not fitted. A diesel engine does not work expansively, as the same volume of gas is present at exhaust opening as at inlet closure; all the expansion occurs in the exhaust. By contrast, a steam engine has a throttle that restricts the amount of steam entering the cylinder, so that by the time the exhaust opens there is little overpressure and no need for a silencer, although you can hear a traction engine at full load bark a mile away if run non-expansively. The biggest advance in steam technology in the nineteenth century was the use of high-pressure steam, working expansively. Unfortunately other inefficiencies, principally its low temperature cycle as compared to an internal combustion engine, meant that it was still less efficient than a non-expansive diesel engine. This deficiency can be addressed by the modification described below.

The high expansion engine consists of a compression cylinder in which the piston travels the full length of the bore, with no headspace (other than that necessary for engineering and expansion tolerances), as in a steam engine. This cylinder is 'oversquare': i.e., having a bore greater than the stroke. For example it could have a four-inch bore and a three-inch stroke. There are two valves (not poppet, as these protrude into the bore) in the head, one controlling an inlet port from atmosphere and the other an exit port which is a transfer port to the expansion cylinder. This second cylinder is 'undersquare', perhaps having a four-inch bore and a six-inch stroke, by way of example; the exact ratios are not important.

We thus have an engine in which the volume of gas at exhaust opening is double that at inlet closure; this means that the gas has been expanded twice as much as for a normal engine. The amount of energy will be the average pressure times the extra stroke, which can be calculated.

The working cycle is as follows: from top dead centre for the compression cylinder the inlet valve opens, until bottom dead centre. The inlet closes, and the air is then compressed until the desired pressure is reached. To align with normal diesel practice this could be when the piston has compressed the air to a ratio of 20:1. The transfer valve then opens, and the air is moved to the expansion cylinder; the piston here has just passed top dead centre, and is beginning to move down. The sum of the two volumes (the decreasing volume left in the compression cylinder and the increasing volume in the expansion cylinder) is constant, so the pressure stays constant for this part of the cycle, the expansion piston having a 'lead' over the compression piston by virtue of an advance in crank angle.

When all the air has left the compression cylinder, valve closure and ignition (or injection) occurs and the piston moves down as in a normal internal combustion engine. However, by virtue of a doubling of the crank throw, it travels twice as far as the compression piston, allowing twice the expansion. The exhaust stroke then follows (no headspace in this cylinder either), slightly lagging the compression piston; that is, rising so as to deliver a fresh charge of air as previously described.

It will be seen that the engine employs a two-stroke cycle, the engine developing twice as many power strokes as a four-stroke engine, which offsets the extra mass of the compression cylinder.

There is a further advantage in having no headspace, inasmuch as there is no spent gas to act as a contaminant to the new charge. The principle advantage however is the extraction of more energy from the fuel by expanding it to twice the volume achieved in a normal engine. Throttling could be by varying the fuel supply, as in a normal diesel. More efficiency could be achieved by interposing a heat exchanger, working from exhaust, just after the transfer port.

The actual values for cylinder dimensions and volumetric expansion could of course be varied in a test engine; figures quoted are by way of example only. The engine could be multi cylinder: a horizontally opposed two compression and two expansion cylinder engine would be less than twice as heavy by virtue of the opposed masses balancing each other more nearly, with no need for 'dead mass' balancing. There are further modifications that could be embraced, which, for the sake of clarity and brevity, will not be addressed here.

A TURBO-RAMJET

I wish to discuss a new type of engine, which I will call a turbo-ramjet, as it combines the principles of both.

The primary use is likely to be for supersonic flight, but this does not preclude other uses. The perceived need for the new engine is that ramjets need to attain a high velocity to function, while gas turbine engines have an upper useful limit of around two to three thousand miles per hour. This is principally because the high angle of attack needed at this speed limits the low speed performance. Nobody has as yet proposed a variable pitch turbine, as the mechanical complexity would be expensive and might decrease reliability. Another drawback is that compression ratios are not very high in gas turbines, leading to a lower efficiency if compared to a piston engine.

The main if not the only element of the new engine is a centrifuge, supported on a shaft and with a 'firewall' or heat insulating divider separating it radially into two compartments. Visually it would look a bit like two dustbin lids bolted together at the rim, a hollow bearing running through the centre, and a sheet of insulating material sandwiched between the two, but not reaching the rim.

The method of working is as follows:

1) The centrifuge is spun up, and the air within it is compressed towards the rim due to centrifugal force.

2) Fuel is injected and ignited on one side of the firewall, thus raising the temperature and decreasing the density of the air on that side, giving a pressure differential.

3) More air flows to the combustion side as the pressure drops, forcing the hot exhaust products out of the hollow bearing on the exit side of the centrifuge. The engine is essentially a radial flow turbine.

133

There are some practical details, as follows.

1) Air entering the centrifuge needs to be accelerated, so two or more curved guide vanes would be useful to direct the airflow, which needs to be at the same speed as the centrifuge at the rim, progressing there by a spiral path.

2) Reciprocally the reverse effect is required on the output side.

3) There is a need for a starter motor to run up the centrifuge, and as a method of control to prevent over-run, and as a power take-off; a dynamo or combined dynamotor might be a good idea.

4) High-pressure fuel injection would be provided by the centrifugal force of rotation, but throttling would be required. This would be more easily achieved by reducing the flow into the turbine at a slip ring valve on the bearing rather than at the injection nozzle. Thus the amount of fuel in the fuel lines past the valve should be at a minimum, to avoid control lag.

5) The entry port needs to be rather small compared to a conventional turbine, so a 'ram cone' will be needed at the front end of the entry port to collect as much air as possible.

6) Similarly, a 'rocket bell' would be useful at the exit end, to increase the reaction thrust by directing the gas rearward, using the expansion after it has left the exit. The configuration could be more or less identical for both ends as they perform equal and opposite functions, with the exception that the exhaust will be hotter than, and of greater volume than the input gas.

The operating cycle is similar to any type of internal combustion engine; gas is compressed, a fuel is burnt into it, and then the gas is expanded to a greater degree than it was compressed. The efficiency of this type of engine is largely a function of the compression ratio, thus particular attention needs to be paid to getting the highest possible compression ratio in the centrifuge.

The principal advantages of such an engine are:

1) It would function at any speed from zero mph to its maximum design speed.

2) Its extreme simplicity, having only one moving part, which is not complex or made of a large number of accurately machined separate pieces as in the case of a conventional jet engine. This should give a low product cost and short manufacturing lead-time.

3) Light weight.

4) Potential for a very high compression ratio, giving high thermodynamic efficiency.

5) High reliability, as there is very little to wear, or go wrong.

6) Low running costs as a result of the above.

7) The engine would scale well as compared with a gas turbine, which needs a constant clearance between blades and casing, a problem if scaled down, or compared with a piston engine, which becomes ponderous and slow running if scaled up.

8) A wide variety of fuels could be used.

SECTION 3

PER OTIS AD ASTRA

Here are yet more inventions, this time ones out on the edge of current technology; i.e. we know how to do it, but don't have the materials to hand just yet. There is plenty to do with regard to design however, so that much of the groundwork is done when the materials do come along. It is also written in the first person, which is usually a bad idea!

This article is longer than average. It is a kind of 'model answer', to show how ideas can be developed for an as yet non-existent technology. There is still plenty of work that you can do however, particularly with the maths, which I am not very good at, and also lots of leads and detail to fill in. These are outline proposals; later one creates models. In this article I will show how to create a model in the head, so you can climb inside it and push and pull things about. Eventually you may get something that can be translated into engineering practice. Engineers are the underrated 'makers to happen'. Newton was sometimes criticised by people working later in the Midlands: all this high-faluting mathematics was all very well, but what they wanted was to know how to forge steel, spin cotton, make pottery, and all the other technical requirements for industry.

You may of course find several major errors in this part of the process. I have made, and corrected, half a dozen while writing and there will certainly be some I did not spot. Never worry about errors in the early stages of work; you are bound to make them and it is part of the creative process.

People have suggested outrageous ideas. Somebody once proposed an orbiting clock; apart from the fact that it would be wrong in all but one time-zone, at the opposite side of the world you would see the back of it, and the side at six and nine o clock. Another person proposed a globe of water with dolphins in it. They would of course drown. These are flights of fancy and part of the process. Most will drop out later as a conflict of results occurs.

Later articles are shorter, which means there is more work for you to do. My intention is to set up questions to be answered, quite often in widely different ways according to the individual; many different approaches are far better than one. Regarding the title, for the very young or if you are in some distant bush village in Africa or S America, Otis is the name of a major lift or elevator maker, and the Ad Astra comes from the Royal Air Force motto, Per Ardua ad Astra; by hard work to the stars. A lift is a lot less hard work.

If perchance you have the book in such a remote area, with just the local one room village school, you might be tempted to think 'there is not much of a contribution I will be able to make from here'.

You would be quite wrong. There are lots of local things you could improve. For example, yokes are often used to carry heavy loads. A yoke is a piece of wood, shaped so as to fit comfortably across the shoulders and back of the neck, with a rope on each end, sufficiently clear of one's sides to allow for a couple of buckets of water or other loads to be carried with relative ease. They were in common use in England less than a couple of hundred years ago. As the buckets were often of wood, with wide tops like a half barrel the water could slosh around, so flat pieces of wood were floated to stop this.

For such a device, thousands of years old, it may seem a tall order to make improvements. The best way is to look at problems which may arise in use. For example, the loads may be unequal, such as two pieces of wood, or two rocks. This can be solved by making one or both of the rope attachments capable of being slid in or out along the arm until balance is achieved.

As the device is rather wide, it may be inconvenient along narrow paths, or through doorways. By rotating the arm around its short axis, a different cut out at the centre may be made so that it is comfortable on one shoulder in the fore and aft position. A further improvement might be to have flattened ends so two people can take an end each for heavy loads. Extension pieces could be fitted if the original was rather short. Sliding the load back and forth could adjust the weight on each end if the two people are of different strengths. By having one end bent up one could also adjust for different heights or sloping ground if the position of the bent end could be adjusted.

Actually, when I was in Africa, in Mukumu province in Kenya, yokes were not used. The girls used to carry four gallon (forty pound), ex-paraffin cans of water on their head, with a little grass ring. Even very young children did this, although maybe they did not fill them all the way. Clay pots seem to have been discontinued. This gave a very good stance, plus a pronounced lordosis (the lower spine curvature) which is also a good thing, but it may have given neck and back problems later, particularly as they used to carry sacks of maize and potatoes, which are heavier than the water. Old men sometimes put their stick across their shoulder, and draped their arms over it, so they had got the idea, but men never carried things.

We now have a multipurpose tool, better than the original.

Again, a lot of time is spent in planting seeds in a rural economy. Maize is often planted one seed at a time in rows a foot or more apart. You could invent a seed planter consisting of a tube with a press button at the top, which releases one seed at a time. An adjustable plate an inch or so up from the bottom would prevent too deep a penetration, and the seeds would be planted at the correct depth if the end was pushed into the ground. A slot cut near the bottom would show when seeds were running out and another hundred or so could be slid down the tube. Planting would be faster, more accurate, and a lot of back breaking bending avoided. You will need to do some detailed work on the seed release device. There are lots of other examples of such simple improvements to be found if you look round at what people are doing, and what tools they are using.

So much for the preamble, now for the main story!

The other day I decided to build a skyhook, as one does. This is a cable suspended from a space station in a synchronous orbit at twenty two thousand miles up, a la Arthur C Clark who first proposed the idea in a science fiction story. As this kind of endeavour tends to take a lot of time and money, neither of which I have much of, I decided to build a virtual model to test out the principle and look for problems that might need fixing.

Obviously one would need rather strong string. One only has to look at the thickness of suspension cables on a bridge to realise that steel won't do; steel cable can support about a hundred tons per square inch, and that would not be nearly enough. In fact twin suspension cables don't seem to be used so much for bridges nowadays. Instead they use separate cables from towers at each end at intervals down the bridge. This allows for thinner cables and spreads the load, but increases the longitudinal compression load on the deck, which can be partly alleviated when the two ends meet by pulling them together. There is still a limit however, for the further the reach of the cables the more acute the angle they make with the deck and the greater the tension.

One can of course make the towers higher, which makes the cables longer and heavier and eventually the thing does not scale. Not to the person footing the bill anyway. There has been talk of using Kevlar, or some other lightweight organic fibre for very long bridges, but I don't think anyone would dare without building smaller ones first, and doing a lot of materials testing down the line. I expect they will use it in lifts first; if you go up in a lift in a very tall building you are likely to change lifts halfway, for otherwise the cable weight and elastic bounce would be excessive. If our brand new Kevlar cable breaks in a lift, there is only the inconvenience of waiting until they let you out, for the braking system prevents a fall. Bridge collapses are rather more catastrophic.

I decided to use carbon fibre cable, a sort of Buckytube like a tubular sheet of graphene. I did not want to twist it like a normal rope, as this reduces the strength (if you look at a twisted rope, on the outside of the turn the fibres are going at near right angles to the load.) The reason for the twist is to capture loose ends and to make it easier to coil; also of course rope is made from a number of short fibres To contain my microscopic strands I enclosed them in a bigger tube, and then again, to get something easier to handle. I am not sure if this material is readily available, but I ordered something capable of standing a thousand tons per square inch load, just one order of magnitude greater than steel. The technology can catch up later if it needs to.

To test my cable I decided to build a tower. To build one tall enough in reality would be impossible of course, but within a virtual model one can embrace impossible things, to test ideas, so long as they are taken out afterwards, when one is left with a set of workable proposals. It is rather like using a leaky argument to cross a river to bring a serviceable boat back so that one may then cross in safety, the leaky argument having sunk without trace in meanwhile.

To hoist my cable I used a kind of vertical rack railway, with a hundred horsepower electric motor fed from the ground. There would be no point in adding the weight of a diesel engine, which would run out of air a few miles up anyway. The density of my material was about 2, ignoring voids, so it weighed about a pound a foot for a one-inch cross section cable. I realised that I could run into trouble at around the five hundred-mile mark, as the weight of cable would be near the breaking strain. Rope at the chandlers is usually measured by circumference; I make this point in case anyone does build a real one and gets confused between a six inch cable that is measured round the circumference, when they really want one that is six inches thick. They confused feet and metres once at NASA, and lost a probe.

At first, hauling my cable was easy, but the load increased until I was hauling nearly a thousand tons. One horsepower is about a quarter of a ton per foot/second, so my rate of climb was only about a forty seconds per foot. With twenty two thousand miles to go, I was going to have a long wait, and also I was getting a voltage drop at my motor. A twenty-two thousand-mile power line seemed a trifle impractical, particularly as I was going to need a few hundred thousand horsepower to pull my increasingly heavy cable.

While I was pondering these problems, the cable broke. I forgot to mention that my cable stretched rather a lot; about ten per cent at full load, which added up to an energy density of about fifty foot tons per pound, quite a lot better than dynamite. Once broken, the severed end accelerated down at about ten times the acceleration of a shell down a gun barrel, while the other turned my hoisting gear into a splash of liquid metal. Had it been attached to a space station it would have been history. I decided that there would have to be an emergency decoupling mechanism for a real space station; the cable would be shed the moment the load decreased beneath a threshold value, and the station would fly off at a tangent until it had re-established a higher orbit. There would still be the problem of the upward recoil above the break; the shockwave would blow the end of the cable off at several tens of thousands of miles per hour. I strengthened the top few miles, so that if there were a break, it would not be there; by the time the shrapnel came hurtling past, the station would have moved out of the way. The cable expanded laterally as the load came off, giving a shock wave travelling down the cable, made visible as the cable glowed white hot with internal friction as it passed. It also bounced; that is, contracted again after reaching full lateral extension, and it was not long before bounces and reflected bounces got out of step, and weakened at high temperature, it started to tear apart. On reaching the atmosphere it burned up giving a spectacular fireworks display, enhanced as bits of the tower, severed into several pieces by the flailing cable, made a more stately cometary descent. There was quite a hole in the ground too, as the bottom end of the cable poured into the ground at a few thousand miles per hour, like molten lead into styrofoam.

A bit of a redesign was needed. For my second attempt I decided to add length from the top of the cable, and double up the thickness every two hundred and fifty miles so that the cable never went over fifty per cent breaking -strain. There would also be an imposed load; that is; the cargo on the way up, and another acceleration load. These would be minimal as a percentage of the total cable load however. I also put in a thousand horsepower winding motor every hundred miles, and a lot of solar panels as I got higher.

Things went well for a while, but doubling up every two hundred and fifty miles meant that I had five hundred tons for the first stretch, then another thousand and a half for the next two fifty, then another three and a half thousand, and so on. Five and a half thousand tons for the first seven hundred and fifty miles! I realised that I could use less rope by having a continuous taper rather than increasing in steps, which about halved my problem, but it was still rather a lot.

I went back to my suppliers and asked for something much less elastic and stronger. They came back with something with partial cross-banding, more like a diamond structure than graphene, which was better, but also heavier. It was still four times as strong per unit weight as before though. The stored energy was reduced under tension; but still large; it was a problem that was not going to go away. Breakage was not an option. Now I could double up every two thousand miles. Also the inverse square law was working in my favour, for at four thousand miles gravity was only a quarter that of the surface of the Earth. The next eighteen thousand miles did not add a lot of weight for the mass involved, particularly as centrifugal force was becoming an increasing help, being equal to gravity at 22 thousand miles. The force of gravity is about one foot per second, I think, at that height, but check my figures.

Then it got struck by lightning. This is only to be expected if you fly a conducting string in a thunderstorm. Unfortunately, lightning tends to occur in the lower atmosphere where there is oxygen about, and carbon burns. Luckily it was quite a small bolt and it was raining quite well at the time, otherwise I could have burnt out again. The cross bonding in the string prevented a runaway detensioning, so it protected against further damage from a few strands breaking.

I remember when young having a seen a faint dark smoke flickering above the elm trees in the field opposite, in sultry weather before a thunderstorm. I later realised that this was caused by gnats and other insect being swept up in the static discharge above the trees. Perhaps they also prevented the trees from being struck, for a lightning conductor does not only channel a strike, it sprays off electrons and reduces the potential, reducing the probability of a strike also. Or maybe they used the field for dispersal; flying is a lot easier if you have negative weight, and unavoidable if you let go of your leaf. Or maybe they just wanted to dance. On still summer evenings you often see a cloud of dancing gnats in the garden, usually under some overhanging bough. They always choose the same spot and I believe they respond to a static field. If you have a spare Van De Graff generator or an old Wimshurst machine, you could try it out on them. A party balloon stroked with a dry handkerchief might do. Nobody seems to have done any work on the response of insects to electrostatic field, or electromagnetic radiation, which may set up weak standing fields. Maybe social insects, like ants and bees, which have to navigate back home, have a kind of 'field map' in their heads. It might explain why honeybees get lost nowadays, due to all the cell phone masts, leading to colony collapse. Many insects use magnetic fields as well as having a sun compass, so we are halfway there. Perhaps we could give them a homing beacon.

Field is also important in the development of an organism. Leaves have a field, which can be photographed; cut a hole in the leaf, and the field will still be projected into the hole. Maybe structure is to a degree conditioned by field, the growing plant or animal projecting a field, then growing into it. After all, growth is only one aspect; there must also be limits to growth. The cells in your finger have to know when they have attained the required thickness; otherwise you might end up somewhat blobby. Cancer cells don't know when to stop; maybe they could be controlled by an imposed field? At least there is a new area to explore, with not much work done so far, always a good sign if you want to make your mark.

My lightning protection consisted of throwing out charged graphite particles to discharge the static field between the cloud and the wire. This probably only enhanced an effect which would have occurred anyway, for although a lightning conductor will take a bolt safely to ground, most of the time it gives a point discharge and runs the static field down. As belt and braces I set up a kind of umbrella rib affair, with copper wires dangling to ground tethers, splayed out to make a kind of electrostatic tent through the worst of the atmosphere. Above a thunderstorm there is lightning going off up into space. This was a surprise discovery, although it could have been predicted, for when the charge on a cloud suddenly collapses, there should be a kind of Newtonian reaction, with things, possibly protons or charged atoms, going the other way. It is much more diffuse than a bolt of lightning, spreading in an upward fan, as it has nothing to home on to. Things might be different if a conducting cable were put in the middle of it, so I decided to keep my eye on it with no action for the time being.

I had included sensors in the build; one sort to cover a wide range of sound frequencies and another to measure current. There seemed to be all sorts of stray currents in my cable, probably due to the sun's coronal discharge, Also, the wire was singing You might think any note emitted by a twenty two thousand mile violin string would be rather low, but there were overtones aided by the very high tension. The moving lift cage or pod created waves as well; there was a lot of modelling to do to ensure that all these various running waves did not add up at one point to cause an overload. I put in dampers to reduce the amplitude at intervals, large slow moving masses travelling long distances at right angles to the cable for the low notes and smaller stuff for the high ones.

I was also getting sharp reports every so often, and realised that very small items were impacting my cable. I guessed it was either cosmic dust or bits produced from our efforts at dumping a few thousand tons of rubbish into space. There were two considerations; first, although slight, there was a long-term erosion problem, and secondly, where there were little bits there would be bigger bits as well. Small stuff could be shielded against; the solar panels would do it where fitted. A small puncture would be produced, with a cone shaped spray of gas, which would do no damage at a few tens of metres. Most stuff is in low earth orbit, around a couple of hundred miles up; the lower the orbit, the higher the speed. A few tons of spacecraft, redundant or otherwise impacting the cable at five miles per second would sever it, particularly as the lower end of the cable is the thinnest. An electrostatic shield or a thin membrane might deflect dust, but bigger stuff would have to be intercepted. A battery of lasers could vaporise nuts and bolts or the splinters from exploding ones used to separate components. At five miles per second you would need to act at a few hundred miles out, so a good radar system is required. Big stuff would normally be of a known orbit, but orbits can drift so one would have to plot them a few years ahead, and send someone to steer them to a safer course, or dump them in the ocean if they looked like getting close to the wire. We could start to think of a clean-up satellite, to start to reduce the amount of rubbish up there and send it down as a firework display. We could call it Lucifer.

Extra-terrestrial objects like the recent Russian meteorite are very rare, and unstoppable with present technology. The chance of one hitting a thin wire when one only hits the much bigger target of the earth rarely are vanishingly small, so we can ignore this problem. Breakage for the wire is not an option however, so we should keep the problem in mind and fit the counter technology when developed. Should the wire break, it would flail most of the way round the earth in bits or otherwise, and even though the ground tether would be near the equator, there is still a lot of land it could impact. Also of course we could lose the space station.

We would need a bit more than 22,000 miles of cable, for its weight would pull the station down to a lower orbit, where it would speed up and lead ground zero. This would lead to it wrapping itself round the earth like cotton on a bobbin. Although gentle at the tether, the cable's descent would get faster and faster down the line, with the station impacting at between five and six miles per second, equivalent to a moderately slow meteorite. Putting it a bit higher, or sending another wire up higher with a counter weight like a conker on a string would do the trick, but it all adds up to more mass, all of which has to be sent up by rocket, thousands and thousands of tons of it. Also, the wire would not be straight, following instead a graceful arc, the parameters of which would depend on the masses of the station and counterweight and length of wire.

Having got all this mass up there, we have to get one end down again, and, counter-intuitively, it will cost just as much energy to get it down as it did to get it up, for we cannot just dangle the cable; it would merely float alongside the space station. We would have to put a rocket at the end to slow it down. It wouldn't actually slow down at first, but go down and try to get ahead, tugging on the station. We could retrieve some energy as it spooled out, via a generator, rather than creating a lot of waste heat to be disposed of. This energy could be used to power an ion rocket at the end, which would save lugging a lot more rocket fuel up. We would have good solar arrays for electricity in any case to make up the deficit.

Next, I decided to give it a test run. The passenger pod was a bit like that of a modern airliner, but on several floors rather than long and thin laterally. Cargo goes unpressurised unless there is some reason why not. Once again, there is the winding problem; it takes a lot of energy, to loft a mass straight up. Your airliner may climb at four hundred miles per-hour, but the angle is very unlikely to be much more than one in five, which reduces the lift to eighty miles per hour. Also, running cables through pulleys at high speed is a bit nerve wracking, so I used banks of induction motors, the pod coasting in between, and the spacing increasing as one got further out. One big boost with a deceleration at the end might another option.

I decided an average speed of two thousand miles an hour would be a reasonable target, maybe starting at less for test runs in case of teething trouble. This still gave a flight time of 11 hours, so it would be a bit more like a trip on a plane rather than a rocket. If you are in a hurry, go by rocket. It will cost a bit more though!

We would also have to consider the lateral thrust on the cable, for as the pod rises up the cable, it has to gain speed to keep it in line between the station and the earth. This would bow the cable, which could be kept within acceptable limits taking into account the extra load and the energy required to accelerate it. The nub of the problem is having something travelling at a couple of thousand miles per hour a few inches from a static cable. The slightest hiccough in a magnet winding might give contact, with unwelcome consequences, plus a lot of plasma.

The 'escape velocity' for the earth is 7 miles per second. You can of course escape at any velocity you like; what is meant is the instantaneous start velocity, or, to put another way, the impact speed of an object falling from infinity. This trades off against orbital height. At our target height of 22 thousand miles it is only a bit over three quarters of a mile per second. There is no cheap route off the planet! For convenience you can think of the work needed to achieve orbit as coming in two fractions: first, the energy needed to raise a mass to the required height, and second, the energy required to attain orbital speed. The higher you go the greater the first and the less the second, until at infinity you have zero speed, all the energy being in the height above ground. We will of course gain some energy as we send things down, via our linear motors running in reverse. However, we are likely to be sending more up, by way of moon and Mars dwellings, and big orbiters around the distant planets, to say nothing of interstellar probes, with heavy nuclear reactors. We might even send thousand ton ice cubes to Mars, if we want a nice little lake there. There is a lot of potential payload in the next hundred years, and a lot of payload means a lot of pay. Also of course, once it was established at the cheapest way to go everybody might want one.

There is little competition. Rockets require about a hundred tons of fuel for every couple of tons lifted to low orbit. A rocket is inherently inefficient; you only have to look at the exhaust temperature to see that. Also the first 'g' developed is wasted. Accelerations must be low for human beings, and one of them is used up merely balancing gravity. There might be a case for 'sprint' rockets for cargo, but then everything has to be made stronger for the extra loads imposed. Although we need a lot of horsepower hours to get up to 22 thousand miles, horsepower hours are cheap and there is plenty of solar power on the way up. We could use banks of very thin condensers to store power at night, and night gets a bit shorter the higher you go, or at least the umbra or total shadow part does.

We of do not of course need to put the space station at 22,000 miles, since the whole thing is geostationary, with a bit of leeway regarding ground zero because of the curve in the cable. We could put a viewing platform for tourists at fifty or a hundred miles, so anybody could go into space, for a consideration, of course! Or we could if we liked put it at 240,000 miles; much more and we might get in the way of the moon as it would intersect our orbit on occasion. At this height it would be describing a circle of approximately one and a half million miles circumference in twenty-four hours; 62,500 miles per hour. This is a very respectable speed: Mars in twenty days! Or a bit more if you consider the slowing due to climbing the residual gravity well of the earth moon couple, and the sun's gravity-well out to Mars. And all this for free, as the energy required would be taken from the spin of the earth. A super slingshot! Don't worry about the earth slowing down though; millions of times more energy is robbed by the tides, with little effect, the earth is so massive. The earth does slow down a little, in the order of milliseconds per century, while the moon is pumped up a few centimetres higher per year, due to the slingshot effect of the tidal wave lagging the moon's ground zero.

There are a couple of slight problems with going to Mars though. Firstly, we are pointing the wrong way! The earth's equator is tilted at about 23 degrees. However there are two intersection points where a projection of the plane of the equator intersects Mars' orbit, as it does for all the rest of the outer planets. We will have to find some other way of getting to Venus or Mercury, probably a backswing to reduce orbital velocity and then drop down to a lower orbit, with a bit of rocket correction to reduce the ellipticity. Or we could perform a loop round Mars to head off in the required direction.

The second problem is that we have to stop when we get there, otherwise it will be a case of hello and goodbye as we sail past. We could use rockets, but the payload will be immense. No problem. Mars colony will assemble a twin geostationary orbiter to the one on earth. As our interplanetary shot latched onto the Mars slingshot, an equivalent mass would be detached so there was no Newtonian reaction to send our slingshot sideways. We could make up a useful package, or just send off a lump of rock, with perhaps small thrusters to correct the orbit so that it did not come back to haunt us when least wanted, for it will be in an orbit that will eventually intersect. We still have a problem however, for our space package is still travelling at the same speed, but now in a circle instead of a straight line. We cannot just wind it down to the centre of our slingshot device however; this is actually impossible, for if we try, we will add energy to the spin which increases the centrifugal force and also moves the centre of gravity, so that by the time we have got to the centre it has moved on, and again and again. It would actually take an infinite amount of energy to get there, which is one of the reasons why black holes cannot work. If you do my kind of maths you will realise that although the energy approaches infinity, the speed approaches the square root of infinity, which is still infinity and you cannot get there anyway, so it hardly matters. To wind our package down we will send an equal mass up, counterbalanced either by pulley or a pair of linear dynamotors. This conveniently provides the next reaction mass for the next consignment.

Luckily the spin and equatorial tilt of Mars are more or less identical to Earth's, even though there is no big moon to stabilise them. This is odd; the chances of both parameters being the same are astronomically low. Maybe the Martians came here and made the place a bit more like home, and then regressed, what with having all those nice trees to climb.

Eventually, given the limitations of planetary slingshots, we will build one of our own. A nice empty bit of space is required; a LaGrange point might do, provided a lot of other bits and pieces have not collected there. The earth-moon system has a couple of interesting ones, leading and lagging its orbit. Here the gravitational fields of the earth, moon, and sun balance, locking in any object placed there.

To make our slingshot we would need two contra-rotating masses; a couple of thousand tons might do, with an electric motor of say ten thousand horse-power to spin one against the other. To power this we will need a big solar farm, or a nuclear power station. Actually both would fit well together. We will need a large array to radiate away the waste heat from the power station, so why not put it back to back with a large solar collection array? As a large array would experience considerable thrust from the solar wind and light pressure, the heat radiator would partly cancel this. A small ion rocket could make up the difference as required, although the small displacement from the centre of the LaGrange point would also help, so long as the thrust was not so much as to push us out. We could spin at one revolution per minute instead of once per day. If we used an orbit of only 500 miles around the contra rotating masses, this would give us a speed of roughly 186,000 miles per hour. We could reach the nearest star in 14,400 years, if we had not forgotten or overtaken our probe with better technology by then.

One problem would be providing the colossal amount of energy required. To find this we can use the formula: energy = half the mass times speed squared. I don't altogether approve of using formulae without understanding what they mean. It is of course a very quick way of doing things, but when the results don't work out as expected one can be rather at a loss. I use energy in foot pounds, because I was brought up that way and America does too. There are metric conversion tables if you prefer metric.

A foot pound is that amount of energy required to raise one pound one foot. If the pound is dropped back one foot, its energy of movement, or kinetic energy, will also be one foot pound. The amount of energy required to double its speed is however four times as much. How so? Well, if we drop one pound for a time of one second, it will end up travelling at 32 ft per sec. That is because that is what the earth does; it is of such a mass that its gravitational field will impart that speed. Try it on the moon and you will find things drop at only a bit over five feet per second, because the moon is a lot smaller. As it started from rest, the average speed will be only 16 ft per second (hence the half in the formula) Supposing we now drop it for two seconds, it will fall 16 feet in the first second, as before, but during the next second it will start at 32 feet per second and end up at 64 feet per second, travelling a total of 48 feet, which plus the 16 feet in the first second is 64 feet, which is four times as much as in the first second and four times the energy; hence the squared in the formula. I hope that explains the effect.

Supposing our probe weighs two grams or a thousandth of a pound; at 186,000 mph we would need. 186.000 X 5000 squared (to bring to feet per second), divided by 32 (I have shortened the half over 64 bit of the equation) to give the kinetic energy. We can then divide by 550 to get horse power seconds and divide again by 10,000 because we have got ten thousand horsepower, then divide by 60 to get minutes and again to get hours. We might as well divide by 24 to get days as well. Off the top of my head, that is about one and a half days to get one thousandth of a pound up to speed. A real probe might well weigh several pounds including some shielding despite micro-miniaturisation, so we have a three year spin up. This means we will have to settle for some lesser speed, although substantial.

It gets worse, of course; to prevent the station going in the opposite direction to our probe we would have to fire a reaction mass in the opposite direction, doubling the required energy. This would become more important for larger masses at lower speeds. Also of course there is the problem of accelerating the mass of the cable up to speed, although this would be a one-off, as the cable would stay in place at the required velocity after our probe had departed.

Actually, the centrifugal force would be a serious consideration long before we had reached such a high speed. I use the example to illustrate the exceptional energy required to attain high velocities.

The formula to find centrifugal force is: centrifugal force = velocity squared divided by radius. You will probably find centrifugal force called centripetal force (which it is, because the force acts towards the centre to stop things flying off) in many books. I use the common parlance however.

You can work out your own results; to explain in rough terms what is happening, imagine a clock face. At 12 o'clock is a mass, travelling clockwise at a velocity of whatever you choose. By the time it has got to six o'clock it will be travelling at the same velocity in the opposite direction, so the centrifugal force is that force required to reverse the direction of the mass, in whatever time it takes to make a half turn. My explanation is a bit simpler than the formula in that you do not need to know the radius, although of course it is there, embraced in the speed of the mass and the time to get round, which determines the radius of the circle. You also need to allow for the fact that the mass has traversed the clock face from top to bottom as well.

There are more considerations before we start to build our Skylift however. The first is where to put it, or at least the bottom end. With the prospect of thousands of tons of cable pouring out of the sky at meteoric speed, no country in the fall zone will want it.

We could of course have two or more tethers, each capable of taking the load with ease. This would give the advantage that we could sling things between them. Also, two or more countries could be connected. There is an advantage in more than one connection, particularly if the excess load occasioned by breakage of one could be taken up by the rest. We could offer free lifts to those countries affected, but you have to get the agreement of all, and the last might hold out for an excessive payment, which is not fair, so the rest will want the same. Fundamentalist groups might object, calling it the Cable of Babel. You could explain that the proliferation of languages is a good thing, so they should support it. You could offer to make their language the one used on the Internet as a bribe; it will cost nothing. Even with an agreement to build, we still need a site. A floating platform has its attractions, for it could drift slightly to allow for the drift at the top due to varying load going up and down. We could have a heavy cylinder, several hundred feet deep, immersed in the ocean, attaining neutral buoyancy when the top few feet were exposed; this would compensate for the varying tension due to load and acceleration of that load. However we have to put things on ships to use it, and we are starting at the lowest possible point, so a ground base with road or rail access might be best. For a ground base, the higher the better, for a mile saved at the bottom translates to a mile saved at the top, where the cable is much thicker and a lot more expensive with the cost of getting it there. We could build a tower perhaps a mile high using steel, or maybe hundreds of miles as some people have suggested, using carbon fibre, and choose some land at over two miles high. With a tower the base should be thicker than the top, for a cable the reverse holds true. There is a trade-off to be worked out. It seems a pity we have to use thousands of miles of heavy cable just to support the last thin bit, where most of the gravity is.

We could ask the question; do we need to put it at the equator? For example, if we had a floating base, we could at first have it on the equator in the middle of the Atlantic, so as to capture the tether, and then head north, paying out a bit more cable as we go. The angle made by the cable would tend to drag the space station sideways to the north, but it would still be in a geostationary orbit. The space station would try to orbit the centre of the earth, because that's where gravity appears to act from. The lateral force exerted by the cable would displace this north, and a break would result in a spiral around the globe rather than an equatorial line. This of course might be worse, embracing more countries. There is another option; if the cable does break, we could immediately make another break at say a couple of hundred miles up. This would mean that only the bottom bit would fall, the rest staying in orbit for a long enough time to do something about it. Another advantage of going north is that at the pole the station does not have to be geostationary. There would be an atmospheric drag if we vary the speed however, so a little rocket thrust would be needed to counteract this. We would of course need two tethers if we stray far from the equator.

So, will such a cable be built? 'Not in my backyard' some might say. Probably not this century I would guess, but if you had asked me about getting a man on the moon a century ago I would probably have given the same answer. There are a lot of technical problems to solve, not least being strong enough string. We do not have to go all the way though; we could start off from one station in orbit to another lower down, maybe a few miles to start with , then a thousand or two. The orbital speeds of the stations would be different, the higher the slower, but a cable connection should give an average, with some tension on the cable. We might extend this idea by having the shortest possible rocket lift combined with the rest of the way by cable. Also, a slingshot in space might be a more attractive proposition to start with.

You will probably find a lot more on the Internet. I have not looked as I find that a clutter of other people's ideas in my head inhibits new ones. First, idea generation, and then synthesis. I have been told there are at least two books on the subject. I have not read them, but may do so later. If I had read up before writing I would have immediately realised that my efforts were of rank amateur status compared with the experts, and not written it.

There is just one small problem. How do we get a few thousand tons of stuff up there? Don't worry, this is all dealt with under 'The Biggest Gun in the World!'

THE BIGGEST GUN IN THE
WORLD. BOOM, BOOM!

Every so often the idea of launching things into space by way of a gun crops up again. H. G. Wells launched his moon travellers there, in a padded shell. A padded cell would have been more appropriate. Even if the barrel had been long enough not to flatten them on launch, the impact certainly would have. He did not appreciate the zero gravity effect of free fall either.

However, Ball, a ballistics expert, did do some tests with a couple of U.S. Navy 16" barrels bolted together, but the rocket program was favoured and funding stopped. Saddam Hussein became a sponsor, but Ball was murdered by Mossad, as Israel thought the huge gun being assembled was a weapon, which it could perhaps have been, but a ridiculously heavy and immobile one.

The design was similar to the German one assembled, but not fired, towards the end of the Second World War, which was a long tube with multiple breeches down the first part of its length. It was only about 6" bore, firing a 150 lb shell, but multiple barrels were intended to deliver several shells a minute on London, non-stop, which would have slowly reduced areas to rubble. They had some experience, bombarding Paris at a range of seventy miles in the First World War, using a 6 inch barrel sleeved into a twelve inch one. At the end of the War Krupps hid it up a chimney but I don't think it was used again.

Teething problems were never sorted out for their new gun; there was a problem with getting the multiple breeches to fire on time, and doubtless barrel erosion was a problem, or would have been after a short time. The trouble with very high speed sliding contact between the shell and the barrel is that plasma can form, despite using low friction alloys. If that doesn't happen, then flame erosion will soon wear the barrel. A new high velocity tank gun a few decades ago only lasted twelve rounds before having to have its barrel replaced. They flame sprayed the barrel with a chrome alloy to make it last a bit longer. Cone boring is common for high velocity guns, more to compensate for the erosion of the driving bands; the barrel gets slightly narrower towards the muzzle; as the bands wear down the shell is still a tight fit.

There have been big guns in the past; I saw one when young in the Tower of London. As far as I remember it was made out of two eight foot bronze sections that screwed together with a bore of about a yard, and at the siege of Addis Abba the German engineers cast a hundred ton cannon, which burst at the first firing. In practice 18 inch seems to have been the practical maximum. Japan built two super dreadnoughts during the last war with 18" barrels, and I was told there was still an 18 inch coastal defence gun at Dover, a while back. Nothing bigger was needed anyway.

The problem is that guns don't scale very well: if you wish to double the bore, you will increase the throw weight eight times. As you will probably need to double the range as well, you will need to make the shell go twice as fast, which takes four times the powder, as the kinetic energy goes up as to the square of the speed. So you need thirty-four times as much gunpowder. Double it again and you run into severe logistical problems. Saddam's gun was about a metre bore, although they did build a test one at half the bore to start with.

There is a convention amongst gun makers: the shell weight is around half the cube of the bore in inches, to give the weight in pounds. For example, a four inch gun as used by the British Navy fires a thirty two pound shell, whereas the American Navy favoured five inches and a sixty two pound shell. A ten inch bore, as used on some of the German pocket battleships fired a five hundred pound shell, while the 12 inch bore favoured by the British Navy in the First World War threw an eight hundred pound shell.

Later, fifteen and sixteen inch bores were used, though the King George the V class only had 14 inch, mainly due to Britain sticking to the nineteen thirties convention restricting battleship size, while everybody else didn't. The Vanguard, finished too late for the war, was 16 inch.

For our super gun I propose a bore of ten feet. If this is a bit too ambitious for you, you can scale it down. Using the armourer's convention, our shell should weigh about three hundred and eighty tons. Actually, as it is likely to be magnesium alloy rather than steel, this comes down to somewhere near a hundred tons. In addition, it will not be solid, but packed with discrete pieces of cargo; however, we might make it longer than the usual one to three ratio, and anyway, a hundred tons is a useful payload! Also, I propose to make it steerable. How so? We will mount it on four old half million ton oil tankers, strapped together, four square. They can sail separately to a tropical coral atoll, near the equator, and it can be assembled there. With a ring of anchors, we can now point it in any direction we like.

One problem with a gun launch to orbit is that the shell is going at maximum speed through the densest part of the atmosphere, which will make it white hot, needing an ablation cone and wasting a lot of energy. For this reason I propose to use a scram shell, a word I have invented to describe a shell with a ramjet in the centre. The front end of the shell will be a flange, an outward cone that chops a hole through the air so that the rest of the shell runs in a vacuum, obviating skin friction. The air is funnelled through the centre of the shell, where it impacts with a fuel; rammed gas carbon would do, or you could perk it up a bit with magnesium dust to get a bit more thrust. The heating caused by friction would self-ignite and not be a loss, as the heated air would be blown out of the rear like a rocket, but one using atmospheric oxygen, thus saving a lot of weight. One would have small rockets also, for orbital correction; for if you launch an elliptic orbit from a point on the surface of the earth, it will try to return there, except that a lot of planet will get in the way first.

What will we put as our hundred ton payload? Anything we like, except people; fuel for the Mars mission, or large prefabricated pieces of orbiters, for this and other worlds. People can go up by rocket to give a more gentle acceleration, but people are light; it is getting hundreds of tons of constructional material up that costs.

What would we use for gunpowder? We would probably not use any, returning to the air gun principle of a compressed gas. Hydrogen would be the best to use, as it is light and travels down the barrel faster. For any gun, there is a limit velocity. This may be increased for armour piercing shells by using a 'sabot' or boot. Thus a slim dart may be fired from a normal barrel, with a washer to fill the bore.

However there is a limit. If you fire a gun with a blank, the gas will issue at a finite velocity, which may be thirty thousand feet per second, but you will never fire anything out of the gun faster than that.

Since we have a lot of room on our four oil tankers, we can store lots of hydrogen. This could pass through a heated sintered block to give it a bit more go and offset the drop in temperature on decompression. We could even have a dump valve near the end of the barrel, such as is used on submarines to re-ingest the air used to blow the torpedo out of the tube, so that bubbles on the surface do not reveal its position. This would save hydrogen and reduce flame - off as it hit the air. In any case, we are not aiming for orbital velocity at the muzzle; we will have to do a trade-off between power from the gun, and power from the scram shell. Also, we will have to choose the optimal path for the shell; too high an angle and we will have less atmosphere to go through and less time of thrust, too low and we may never get into orbit.

We need to consider barrel design here also; the thought of a hundred ton shell resting on the lower side of the barrel is not an attractive one. To avoid this, and consequent wear, we will have the shell follow a ballistic trajectory. Imagine that the barrel could be removed, but the shell still travel at the same speed; use a rocket if this helps. The curve formed will be a straight line distorted by gravity; i.e. it will start off being steeply curved downwards, then shallow out as the speed builds. Actually, friction is not such a great problem at the low speed end, so we can have a bit of barrel support there, but further up it will be freely floating with no weight on the barrel. We still have to have some kind of seal however, to prevent blow-by. This could be achieved by a gas seal, not a seal to prevent gas passing, but a seal made of gas ejected by the shell in a backwards direction in the small clearance space between the shell and the barrel. This would also stabilise the shell in its path down the barrel, keeping it centred and preventing chatter.

Next we have to consider rate of fire; we should be able to achieve several launches a day. This would give a scatter of payloads in orbit, but shepherding, or bringing them all together, does not take much energy, and we can do this quite slowly.

So there we have it: the biggest gun in the world. Cheap to orbit, nearly as good as a sky lift, and much less expensive; in fact we will need one to get the bits of the sky lift up there in the first place. Get your slide rules out; it might actually be built - some day, that is.

WHAT ABOUT THE RAILWAYS?

I wish now to discuss something much more practical, and to show how ideas that are commonplace can be altered, modified and evolved to get a greatly improved product.

I will start with a digression, or, more accurately, immediately embark on one after starting, to show the progression of ideas in something much simpler than a railway network, by way of example.

There is a story that James Watt realised that steam could do work while watching the lid of a kettle being forced up by escaping steam. We will use the kettle for our example. Originally people probably heated water by dropping hot stones into an earthenware vessel, for such vessels have a short lifetime if placed directly on a fire. Spalled stones, too small to be of much use for hearth construction, have been found on ancient fire sites, so it is probable that these got their pattern of surface cracks by being dropped, while hot, into water.

The iron pot was a great improvement, as this could be placed directly on the fire. They were probably in common use in some regions five thousand years ago. One of the early Greek philosophers said, 'The flame wears away the iron (of the cooking pot).' About that long ago, iron was the hardest substance known, but even this was not immune to change. He was actually not commenting on cooking, but of the ephemerality of all things, no matter how substantial they seem, but he used an example that everybody would understand. Iron pots then were of beaten sheet iron, not the cannibal variety.

Later, there was the dedicated kettle, with spout, lid and handle, which you did not cook in, standing on a hob beside the fire or on a trivet over it. This was used unmodified on gas stoves when they were invented, later a whistle was added so you knew when it boiled, sometimes with a push-button lid. Later still an electric element was added, so that it had its own inbuilt heater. Another invention was the thermal cut out; if the electric kettle boiled dry, a strip of metal expanded and triggered a compressed spring, which ejected the plug.

159

The next most important development was the 'forgettle', which switched itself off when it had boiled, without ejecting the plug. Also we had flat elements, so that a small quantity of water could be boiled, for a single cup. Then a window, sometimes with a float, was added, so you could see how much water was in there, plus a gauge, usually marked in quarters. The kettle-jug appeared, and then the cordless one, which was more convenient.

A very good tip when inventing gadgets is to make them so they can be used one handed, for while there are not many one handed people about, those with two can hold something else whilst using it. Also make them symmetrical, so left handed people are not at a disadvantage. I have a cordless auto kettle that can rotate three hundred and sixty degrees, with a big gauged window either side, push button lid, indicator light, and a flat element. You can get one with an electronic whistle too! This was perhaps an invention too far, for it had a normal wide bore spout lid for filling, exactly the same as an ordinary kettle, from which it was derived.

You could add a timer if you like. Do not add a radio, mobile phone, or clockwork orange that barks like a dog when you pick it up. This is accretion, not invention.

Railways are a much bigger invention than the kettle. One of the more important differences is that there is an added dimension, called management. Currently in Britain there are three separate elements to management: one responsible for the track, another owning the trains, and another running them. Probably there is little need to split ownership of trains from those running the service. There may be a case for keeping track separate, but the best way to do this would be to make it part of an integrated transport infrastructure, along with roads. This would free up more money and reduce costs. Currently, if you wish to transport people by coach, you have to pay a few hundred pounds a year as a licence fee, depending on its capacity, then you have unlimited mileage, perhaps a couple of hundred thousand miles per year with more than one driver. To give a level playing field, a railway coach should also pay the same amount per year to use the system. To be really fair, boats on inland waterways should pay the same. A Thames steamer has to pay almost as much a year to use a much smaller mileage, and private licence fees are rising faster than inflation, while you have to pay yet another fee if you wish to use the canals as well. Consequently there has been a big drop in boat numbers and a drop in the net balance of payments as people go abroad for their holidays.

I wish here to concentrate more on the physical structure of the railways, as you have to have something to manage first. As improvements are made, then those from within the railways who have been responsible for bringing them in can move on to management, with perhaps a few people from outside who have managed other concerns.

If you are British, after the weather, it is a conversational gambit to complain about the railways. I mentioned this to someone from another EU country, which shall be nameless, and he said 'Huh! You should try our railways!'

So perhaps dissatisfaction is international. Actually, Britain has one of the most efficient railways systems in the world, with the lowest subsidy. The trains are clean, staff invariably helpful and cheerful, and they run on time, mostly. Even when there were problems a few years back, due to a big safety drive, I commonly got back early rather than late, by catching the train before, which had been delayed. Quite how a half-hour service can be three quarters of an hour late I don't know; why not drop a train and re-label? This is what they actually do on our local branch line, and don't tell anyone.

France has invested more: It is said that every Frenchman buys a TGV ticket via taxes, but if he actually wants to travel on it, he has to buy another one. I have been on the TGV, and it is quite impressive. Unfortunately it stops hardly anywhere. I went from Paris to a station near Bordeaux, but had to get off thirty miles past the station I wanted, and then had to wait nearly an hour for a local train to go back, and that took nearly an hour, as it stopped everywhere. Add to this about one hundred and fifty miles on ordinary rail, mostly in the wrong direction, before boarding the TGV. add a thirteen stop tube journey across Paris to get to the right station, and the average speed came down from about one hundred and sixty something miles per hour, to below thirty.

This brings us to our first radical improvement. There is a huge cost in energy and time in stopping a fast train and getting up to speed again; far more time is lost than that actually waiting for passengers to get on and off.

The simplest solution to this problem is glaringly obvious, if counterintuitive. It is that the train should never stop!

As a corollary we can do away with all stations on the main line, or at least have the line bypass the station where possible, as high speed non stopping trains are unpleasant and potentially risky for passengers waiting on platforms. In France they had an announcement telling people to stand well back when a TGV went through, and they meant it! Many more towns and villages could be included, as only a short spur would be required to bring them into the network. New lines could take the most economical route, avoiding expensive development in built up areas. Also, with greater traction, and a better power-weight ratio, much steeper grades could be contemplated, avoiding the need for expensive embankments and cuttings.

The early railways owed a lot to canals, from the huge labour gangs, to legislation that enabled the purchase of long narrow strips of countryside. Canals had to be absolutely flat, apart from locks, so grades of one in a hundred were a great improvement. The mentality persisted and it became a mantra that rail could not climb hills. The early steam trains of course had only two or four driving wheels, which could easily slip, while low power to weight ration meant gradients caused great delay, which could not be regained on the down slope. Fifteen miles uphill at fifteen miles per hour, plus fifteen miles downhill at forty-five miles per hour, does not average thirty miles per hour, although the train might do that on the level.

Nowadays, with thousands instead of hundreds of horsepower, and two or more driving wheels per coach, any reasonable grade could be contemplated, particularly as the kinetic energy of the train could be used to power up them. The height attainable under kinetic power goes up as to the square of the speed. A fifty-foot climb at a hundred and fifty miles per hour will hardly make any difference to the speed of a train, while at thirty miles per hour it will stop it before the summit.

You have probably twigged by now that carriages can of course stop, but only where people want to. There is not much merit in stopping three hundred people just because half a dozen want to get on or off.

The system would be that just before the train, travelling at say one hundred and fifty mph, has bypassed a station, which would be on a spur off the main line, a carriage leaves the station and is caught up by it, a dozen or so miles down the track. Passengers then make their way down the train to get to their destination carriage, which is dropped from the rear at a suitable distance before their destination. This carriage then becomes the one waiting to join the next train. Thus we have a kind of shuttle; every station can have a carriage waiting for passengers, while a high average speed is maintained. The train never stops; only the coaches stop, these being continually added at the front, and dropped off the rear.

The same system could be used to greatly minimise changing trains. Rather than have people disembark at a station and then wait for another train for a different destination, they could move to their destination coach when it joined the train. The train could then split; the two or more parts travelling at constant speed once they had achieved a sufficient separation for points to change. They could then follow different routes. An intelligent ticketing system could reserve seats; a swipe card system could be used, rather like the trams in Bordeaux. Passengers could be kept track of or missing passengers who had managed to get in the wrong half could be paged, by ticket number if not by name. Busy interchanges could have two or more drop coaches rather than one, restaurant cars and staff changes could be switched at the end of a shift. You would not need long trains. Vancouver has a light rail system that is fully automatic, running individual coaches without drivers every ten minutes. They had a strike once, but the system kept running, with honesty boxes in place of the ticket sellers. Nowadays vending machines take the place of ticket sellers in many stations.

We have now got rid of one of the biggest problems in running a railway; that of slow stopping trains using the same line as fast trains. We might save money on upgrading track, for in effect speeds would be nearly doubled on ordinary lines at existing speeds, reducing the need for track improvement to take ever-faster ones. The system would have to be phased in of course, and some small branch feeder lines might never be converted.

Next we can look at density and passenger numbers. Many main lines are running at capacity, quite often with standing room only for some on a few very busy commuter routes. This is very undesirable for an ageing population, and for passenger safety, and prevents the use of refreshment trolleys. The simplest way would be to increase capacity by using double decker trains. You might have to redesign to drop the floor to the axle, perhaps use smaller wheels, and maybe not have full standing headroom over some seats if the walkway through the upper deck needs to take space. If you are good with a pencil and paper, or do cardboard modelling, there are many configurations to be worked out. Remember that the first double-deckers will be for commuter lines, with fairly short seat occupancy, and you will lose under seat space on the lower deck to house quite a lot of gear, such as electric motors, brake cylinders, suspension and the like. Don't forget to leave space for all that shopping, though!

One problem in those countries that developed railways early is that they have low bridges. A long term plan to upgrade bridges, at least on mainline routes, might be worthwhile. Bridge building gives an interesting sidelight on 'railway mentality'. In the early days many crossings had gates manned by a crossing keeper. However, in the country, where road traffic was slight, it was considered hardly worthwhile to employ a crossing keeper, so bridges were built over country lanes. In town however, where there was much more road traffic, a crossing keeper could be busily employed! This is of course perfectly logical, and has resulted in the greatest road congestion where there is the greatest traffic. This may be desirable for the railways, as it slows down the alternative, but it is not part of an integrated system.

Next you can increase the frequency. You may be told that you have 'got to have' a certain time or distance separation for safety reasons. No you haven't! You upgrade the safety. If pigeons obeyed the same air traffic regulations as aircraft, they would have to fly five miles apart. You could take a look at braking distances also. Trams used to have a magnetic brake that clamped to the rail, and this could be used for ordinary rail. There is a limit though; too much braking and you will tear the track up!

The safest railway signalling system ever devised was 'loop signalling'. I saw it in use once, when I was three or four. The engine driver leaned out of his cab and held his arm out; the stationmaster held up an iron ring a bit over a foot across, and the engine driver collected it as he went by. The system was that a train entering a single-track branch had to pick up the loop before proceeding, and there was only one loop. The loop effectively became the key to the track. I am not suggesting that we go back to iron hoops, but you could have a computerised system. The computer would only have one code number for a certain section of line. It can only issue it once, and cannot reissue it until returned by a train leaving a section. The signalling system and its displays could then be in the driver's cab, or for automatic systems, in the computer. If you are nervous, you could kill the power for electric lines for a distance behind the train until the section was clear; after all, if there is not supposed to be a train on the line there is no need for power either. If a train strayed, the driver or automatic system would immediately notice the power loss and could coast to a safer speed or stop until power was restored.

High running speed brings a higher risk in case of derailment. It is rather amazing that you can run at over two hundred miles an hour, kept on two bits of thin steel by a two-inch flange. If we started from scratch we would not dare! It has been said that a two hundred mile and hour train crash would be more like an airline disaster.

Actually, no. If your aircraft undercarriage fails just before take-off, apart from fire risk, you have a very good chance of stepping out unscathed, the reason being that it is the sudden stop that does the damage. An aircraft, or a train, can slide half a mile or more on the level and slow relatively gently.

Let us suppose that your high-speed train has derailed. This is not a disaster yet. We now need to keep it straight and prevent jack-knifing. Also, we need to keep it on the track, and upright, even if not on the rails. A straight railway track is a good place to slow down gently!

To prevent jack-knifing we need to get progressive braking from front to rear, hard braking at the rear, light at the front; this keeps it in tension and pulls out any kinks as they develop. We could consider decoupling coaches not derailed; they could slow to a halt while the derailed section ploughed on. Next we need to keep it on the railway bed. A dished bed with the centre of the track lower, rather than a flat bed, will help here. Luckily, the wheels on one side will have dropped between the rails, and these will tend to keep it contained. There should be a sufficient flat outside the rails so the wheels that side do not drop, and the train does not tip over. We could improve the chances of keeping it contained by having an outboard skirt, just above rail height, which of course will be below rail height after derailment. We could have a curved piece of metal, which would fit over the rail and act as a brake; thus the train is still in contact and contained even it is off its rails. On curves and embankments we could have something similar to a motorway crash barrier.

The next thing is to stop it hitting anything. Some old bridges have central pillars between twin tracks. These must go; all bridges over should be clear span. Next, railway furniture such as points levers, transformers, signalling gantries, pantograph and telephone poles should be at such a distance that our contained derailed train cannot hit them. We now have the problem of which side it has come off. We don't want it overlapping the other line, as there may be a train coming in the other direction at two hundred miles an hour. If that happens, then you do have an airline style smash. Motorways have a barrier down the central reservation, and rail could too, particularly on curves where there is a greater chance of a derailed train overlapping the other track. We also need an alert system, so the derailment is automatically relayed to a control point and power cut. Using an accelerometer that would trip with the vibration occasioned by derailment could do this; a mobile phone link could be used.

Next, we could see if we could get it back on the rails. This is not as crazy as it seems. There have been examples of derailed trains bouncing back on the rails again; there was a case during the war when the resistance derailed a German run train by removing a whole rail. To everybody's surprise, not least the driver's; the train derailed, there being no rail there, but then crashed over the sleepers for the missing section hit the intact rail past that section, remounted and carried on as if nothing had happened! The reason for this was because there was a slight curve in the track, and as they had removed the inside rail, the other side had not derailed, the wheel flange being pushed against the rail by the centrifugal force. When the wheels on the derailed side hit the intact rail, there was not much else they could do but remount, as the spacing was exactly right.

It would be quite possible to design a 'catch', being a metal tray, wide at one end to catch the errant wheel, but sloping up and narrowing to throw the wheel back onto the rail. You would have to have one each side of each rail, every so often and at danger points to allow for all contingencies. If the train has been contained along the track bed and is still straight, it would have not have much option but to be set on its wheels again after hitting such a 'catch'. If you can get hold of some old stock, and a bit of line, you could have great fun with radio-controlled crashes. You could start with your Hornby double O set.

Next we could examine how people use trains, starting with the people who don't. Most station car parks are full during working hours. If you doubled the car park size you would probably still fill it, and sell a lot more tickets for your double decker trains. You could get some stickers saying 'No standing on top deck'. I know we have just said that standing is a bad idea, but think of the money rolling in while you are waiting for more rolling stock! You could borrow your stickers from the bus station, which will be within the complex.

Many people would prefer to travel to work by train, if only they could park. While we are at it we could abolish road tax, and instead put that cost plus that of compulsory minimum insurance on petrol. The trouble with the present system is that a car costs nearly as much whether you run it or not, giving an inducement to use the thing as you still have to pay road tax, insurance, and depreciation if you leave it in the garage. The present system also puts an unfair burden on the low-mileage driver, who pays the same road tax as a commercial traveller doing many thousands of miles each year. As motorway use declines with the costs placed on use, not the cost of possession, we could take over a motorway lane for rail, thus speeding journey times while cutting costs and pollution. The sight of rail commuters overtaking them at three times their speed whilst reading a paper might encourage a few more to leave the car at home.

As there is unlikely to be much space for parking available outside the footprint of the station, we could have double decker stations also. If you go to a major station, like Waterloo or St Pancras and look up, you will see a glass roof sixty or more feet overhead and an awful lot of empty space. In addition there are few dozen acres of line and sidings. If you go to the nearest estate agent and say you have twenty acres of land for sale, right on top of Waterloo station, he will break your arm off, and try to buy it cheap at ten million an acre. So we might as well put a car park on top of most stations that need it. While we are at it, we could put retail outlets there as well. Most big shopping centres are out of town. The main reason is that rates are much cheaper, and it is possible to put in a car park. The reason for the demise of the high street is not only predatory pricing by the big supermarkets, but that small shops pay ten times as much in rates per unit sale and nobody can park within half a mile anyway. We could have a major retail outlet, or a market-style array of small shops, above any station we wanted. Cheap rentals would not only bring revenue where there was none before, but lots of fare paying passengers as well.

This would greatly increase short haul traffic, so we could look at having double decker platforms also. There is a mantra that trains cannot tackle steep slopes. Originally the early railway engineers erred on the side of caution. Engines were light with little grip in the wet, also they were underpowered. Their main competitors were canals, which could not go uphill at all without the monetary, water and time cost of locks. Trains can actually tackle very steep slopes; at fairgrounds they loop the loop. With more power and many more drive wheels greater gradients could be tackled. The real problem is the length of gradient, not its angle. Suppose we put in an upper level; if you want to see one, go to the Portsmouth & Southsea station. The Harbour line rises to an upper level a dozen or more feet above the lower platforms, mainly because there are a couple of roads to cross. Let us suppose we wish to put in another set of short haul commuter platforms above a normal station. We could use a one in ten slope, and rise sixteen feet. A train, approaching the slope at about twenty miles per hour could climb it under inertia alone, without using brakes or engine power It gets much better as you leave; a free twenty miles per hour boost at an acceleration of one tenth of a 'g'. And all this at no cost for every train, forever. It is so good it might be worth raising all platforms at all stations. The actual saving is only about a pint or two of diesel per hundred tons of train, but over the year and over the network it is worth having, although of course the real saving is on space and time.

The received wisdom says that rail should be used for transport of goods, not people. The problem is that goods trains tend to have a lower power to weight ratio, and go slower, thus becoming an obstacle to fast passenger transport. The only real answer is to speed up goods, and perhaps have more dedicated night services for goods only. It would be difficult to mix the two without redesigning containers so as to have a corridor, to allow passengers to find their correct coach. It would be possible however to have 'car trains' where people could take a long train journey, and put their car on board, as there is sufficient width to leave room for a corridor alongside a car. Also, of course a goods train does not have to be half a mile long. Two or three trucks could be sent off as soon as they were ready. They would be given a destination code, which would be read as they passed interrogation points on the system, and automatically be diverted to their correct destination. The information on the train would interact with the signalling system, making it 'intelligent', as they say. If they were fast enough they could fill in the gaps between passenger coaches, even though they lagged. There could be bypass loops where necessary, allowing faster traffic to overtake if need be.

We hear about 'getting heavy goods off the road and onto rail'. It might be worth seeing what else we can get on the rail as well. The advantage of having long thin strips of land is that there are plenty of organisations that need just this. Water companies, oil refiners, cable, telephone, gas, electricity supply, all need to run unbroken pipes or wires for long distances. The legislation that allowed railways to buy long thin strips is still there. There could be a good case for putting everything alongside the railway, and widening the strip of land if need be. It would be worthwhile seeing what else can go down a pipe as well. Nearly everything you buy in a supermarket could be made to fit down a four-inch pipe. At a modest ten miles per hour a hundred thousand cans of beans could be transported! There might be a bit of friction however, but the idea could be looked at for short haul, perhaps to get from a factory to a railway line; the packing machinery could then be next to the container which could go straight on to the rail instead of threading through city streets.

We have come a long way from a horse drawn truck a couple of hundred years ago that ran on wooden boardwalks, later replaced with cast iron plates, which broke, later replaced with wrought iron, which wore badly, then steel rails. It is still a good system with plenty of mileage yet. Improvements do not have to be ultra-high tech, and a lot of them can be self-funding, while the rest could be funded with re-deployment of current expenditure on roads.

Of course, in the far future we might have vacuum tube railways running on magnetic levitation and linear motors. We could float tethered tubes a few hundred feet under oceans. The best place to try this might be the strait of Gibraltar; there has been talk of a tunnel, but it is deep and in an earthquake region. Also, tunnels under water never pay. A train costs as much as a boat; if you go for the boat option you don't have to pay twenty billion for a tunnel! There is quite a current through the straits; the rivers into the Med only supply about a third of the water evaporated, but it is fairly constant and in one direction, so we can put in turbines for power extraction along with our road and rail tubes. There is probably not the potential traffic yet though.

With greater speed and power we could cross the contours of the land, largely avoiding expensive cuttings, tunnels and embankments. If this ever happens the advances will be built on and funded by the continual improvement of the present system. Short term funding is needed to get a system up and running, but in the long term the system should be able to pay its way without subsidy, and this means increased efficiency and de-subsidising the competition also.

Publicity is very important when doing something new. The Vancouver light rail was done on the back of a World Trade Fair. It is very common to have some kind of showpiece at such events. The most prestigious are Olympic venues, and there is always a fair or Games coming up somewhere in the world. There was a non-stopping helically driven railway at Olympia in London in the nineteen twenties; it never caught on, but that is not the point. A showpiece for technical excellence can sell a lot else because it focuses attention. At the Brussels Expo in '61 the centre of attention was the space capsules. Nobody ordered one, but they showed world class engineering.

We could design an attention catching rail system for the next big venue. To really get people talking we could call it a Non-Newtonian or inertialess drive and let people work out for themselves why it is not. Such a system is capable (read not capable) of moving a mass from position 'A' to position 'B', which is completely outside the space occupied by 'A' with no corresponding Newtonian reaction. That is, it should not thrust against the ground or use any other reaction mass to gain forward momentum.

To do this we use a small Ferris wheel, perhaps a couple of dozen feet across. Within it are a number of seat such as are found at ski lifts, in which people may embark and disembark, perhaps having an 'In' sign on one side and an 'Out' on the other.

The wheel will not only go round, but it will progress longitudinally on an overhead track, the bearing being supported by a bogey, or roller skate, running free. The bottom of the wheel will be just above ground level and rotating at such a speed that it appears to be rolling along the ground, although not actually in contact. It will have to be on the level obviously, to keep constant speed, and will follow a circuit around the showground, giving free transport and a view of the ground. As friction will slow it, we will have to give it a push every so often, but this will not obviate its non-Newtonian characteristic, for the push will merely compensate for its loss to friction, which is the other half of the Newtonian equation.

In use, a passenger will step into the wheel, which is rotating quite slowly, with the bottom stationary in relation to the ground as for any rolling wheel. He or she then sits down on the ski lift seat, which as the wheel rotates is carried up to the top, where it is going at twice the average speed of the wheel, then descends again one rotation on. The path it follows is called an asymptote, I think. The passenger can then step out, or proceed for a number of further revolutions.

The net result is that a mass (that of the passenger) has been transported from position 'A' at rest, to position 'B' at rest, and on the same level. Thus no actual net work has been done. More accurately the passenger has borrowed energy from the wheel as the seat rises, but given it all back again as it descends. This is much more efficient than using energy to accelerate a passenger, then wasting the energy in braking, although many electric transport systems, such as the London tube, do actually have regenerative braking, where a slowing train feeds back electricity into the grid, the motors being in effect dynamotors.

There has been no reaction force at the ground, for the wheel does not touch the ground, nor at the support, for that is free-running on a bogey. So where is the catch? Surely it cannot really be a non-Newtonian drive?

Well, we can leave people to work that out for themselves. In private we will realise that what we really have is a mass in constant motion, which is all perfectly Newtonian. We have cheated very slightly by adding and subtracting to the mass, although it will be more or less constant as people get on and off. If we had perfectly balanced people exiting and leaving, the net result would be the same as a weighted wheel with no exchange of masses. So how do we move a mass from 'A' to 'B' with no energy cost? Actually, no energy is required to move a mass in this way provided there is no height difference and we can subtract the kinetic energy put in at one end from that got out at the other. There will be a slight frictional loss to be made up, of course, mainly air resistance, but this will be minimal at two or three miles per hour.

If you want to go faster, use a device like a caterpillar track instead of a wheel; here you can have a longer dwell time where the seats are stationary in relation to the ground, giving a faster speed for the top section. Progressing any distance might be quite sick making however!

We could hang up our tile: 'Railways done!'

MOTO PERPETUO!

Most people have heard of perpetual motion machines, and most people know that they are impossible. However, they seldom know why they are impossible, other than that they will sometimes say that they break the second law of thermodynamics. This is insufficient; our laws conform to a reality, not the other way round.

Therefore, I wish to delve a little more deeply to see what is possible and what is not in the light of current understanding. There is absolutely nothing wrong in looking at 'impossible' problems; they provide a very good learning curve, and many impossible problems have been solved by people who do not have the word in their vocabulary. It was known for a long time that it was impossible to transmit a key for a code securely, until Rivest, Sharmir and Adlerman in the USA described a one way function that made it easy to encrypt a key, but fiendishly difficult to decrypt it. As a footnote, the problem had been solved a few years earlier by Cocks, Williamson and Ellis in the UK, but as they worked for GCHQ it was kept secret. This decision not to exploit the invention probably resulted in a loss of several billion pounds to the UK. It used to be said that the UK was bad at exploiting invention because we were a nation of shopkeepers. Looking at the High Street, apparently we can't even do that anymore now.

We have a similar situation with other codes and cyphers. It is known that the only truly unbreakable cypher is the random number cypher, sometimes called a one-time pad. They are not used much, because distributing keys is difficult and risky on a large scale. We will take a look at an attempt to crack a one-time pad later. Similarly, quantum cryptography is hailed as the coming completely safe method of information transfer, once quantum computers can break very large prime number codes rendering those in current use obsolete. Actually, quantum transmission can be hacked into using virtual photons without actually doing anything to the encrypted message at all, so a whole new set of safeguards will have to be developed to go with it. Also, quantum encryption may be vulnerable to a time lapse effect; quantum bits don't exactly inhabit the same time-frame as we do, so there may be the possibility of receiving a message before it is sent!

First, we must consider the words. Motion can be fairly described, provided we have an observer and something to observe, as an alteration in distance and or bearing between the observer and that observed. There are plenty of examples. The whole universe is in a state of expansion; within this expansion there are orbital motions of stars around a galactic plane, planets around stars and even at the atomic scale all is motion, with electrons in orbits around nuclei, and presumably some internal motion within those nuclei.

Perpetual presents a different problem. Exactly what do we mean by perpetual? An infinite time might be the simplest answer, but then we have to ask, 'whose time?', for time is not a constant; park yourself in an orbit on the event horizon of a black hole and I will say your time has stopped. You will say that the rest of the universe is over in a flash; our experience of time depends on velocity and gravitational field in relation to an observer.

To compound the problem, there are different types of infinities. I might say to you, 'I have just drawn a line that is infinitely long.'

You might say, "No you haven't, I can see where it stops, just at your feet," to which I might reply, "That it is where it starts; it is only infinite in one direction."

There are in fact an infinite number of different types of infinity. Currently there are several models dealing with the end of the universe, and others that deal with a plethora of universes. They are probably all wrong in the long term, at least in detail. The best we can say is that we don't know, and infinities are difficult concepts to deal with; perhaps a mental abstraction may be useful, but that does not impinge greatly on reality. Usually the appearance of unwanted infinities in maths means that your maths has broken down.

So what do people actually mean by a perpetual motion machine? There are historically two main types. The first are devices that merely go on working, neither generating nor degrading their inherent energy, as in a clock that never needs winding and never stops. The other type is supposed to actually generate new energy.

With the advent of thermodynamics, a third type, which reverses the trend to entropy, became a possible field of study.

Regarding the first type of machine, we have perhaps a model in the circulation of electrons round an atomic nucleus. However, as we do not know if the universe will persist forever, we cannot say that it is perpetual. There was a time when electrons were too hot to be bound to nuclei, and there may be one again. Also of course some electrons get crushed into their nuclei in stellar collapse. All other orbits in galactic or planetary systems must eventually decay, for tidal effects, and the headwind of dilute gas, however slight, will eventually take their toll.

The main objection to energy creating schemes lies in the three laws of thermodynamics. Vis:

1) Energy (or matter) cannot be created or destroyed.

2) Energy tends to run down to entropy; that is, it degrades to a less useful state, eventually leading to a uniform temperature for the universe where no machine can extract energy, due to the absence of a differential for it to work across.

3) As the temperature drops order increases, as in the formation of ice crystals as water freezes.

There are some problems here. Firstly, if matter cannot be created or destroyed, then we can't get started. If there was an initial creation event, then one could argue that, given sufficient knowledge, we could replicate it, creating new matter, hopefully slowly, creation of a new universe all at once being a bit disruptive. There is however no known route to this.

Thus the first law starts with an exception, and may end in one for all we know. One problem with 'big crunch' end of universe theories is that they break the second law, in that entropy (or information) is destroyed. This of course negates the third law also.

The best we can do is to say that the three laws are based on observation of the current state of the universe and are empirical, there being no proofs other than the fact that always, without exception, that is the way things work. Virtually all other laws are time symmetric. For example, if one took a video of small particles jiggling about in a liquid, influenced by collision with the molecules of the liquid, there would be no way of telling if it were shown backwards. If you ran the video long enough however, eventually the liquid would cool, or the container would degrade, or some other irreversible event would occur. In other words, entropy seems to define a direction in time, a sort of 'time's arrow'.

First, I wish to examine possible routes to extracting energy from nowhere. Ignoring mechanical devices that use a motor to drive a dynamo, which drives the motor and the like, we can take a closer look at nowhere! Now, you can't make a vacuum out of nothing. To start with, it is virtually impossible to make a completely empty space; even our best vacuums are pretty busy. Pump down to a millionth of an atmosphere and you still have several billion molecules per cubic centimetre. Even if we could get rid of all the gas there would be stray magnetic, electric and gravitational fields that would be difficult to negate. Supposing we could remove all this however, there is still the vacuum energy. Not a lot is known about this, although theoretical calculations show it is there. Unfortunately they also show it should be about one to the forty third times bigger than that actually observed. This is probably the biggest mismatch between theory and experiment that we have. Strangely and perhaps not coincidentally, it is also the ratio between the strengths of electric and magnetic fields, so perhaps we have a route to connecting electromagnetic field and gravity.

However, it is there, as is shown by the Casimir effect. To demonstrate this, arrange two plates to be placed very close together in the best vacuum you can get. You will find a very slight force pushing them together. A large-scale similar effect was noted in the C16th. To paraphrase:

'When two ships are becalmed near each other, in a confused sea after a storm, they slowly come together as if by some magnetic attraction. Eventually the boats have to be got out to pull them apart before they clash together.'

The reason, not understood then, is that ships act as wave reflectors, which means that there is less wave energy in the space between them than outside, resulting in the waves outside pushing them together.

The Casimir effect is exactly the same thing. The vacuum is not empty at the scale of the very small. It is a seething mass of virtual quantum particles, entities that briefly occur as paired opposite charges, which more or less instantly annul each other. You can think of it as being like a very fine froth, or raindrops falling on a puddle. However, in their brief virtual existence they exert a pressure. In the space between our plates there is no room for larger wavelengths, hence the slight pressure exerted on them, there being more energy outside the plates than in the gap in between.

So how could we theoretically extract energy from this effect? With difficulty. Perhaps, for the effect is slight, we will have to expend as much energy pulling our plates apart as we gain from letting them come together, but only if we pull them apart in the same plane.

We can create a model to extract energy from waves at sea to show a principle. To do this we can build a framework on which we mount two horizontal wheels or carousels, each having paddles vertical to the water surface and tangential to the wheel rim, lying side by side, not quite touching. The paddles are capable of being turned ninety degrees horizontally so that they lie out of and parallel to the water surface, which they do through three quadrants of a revolution. As with our boats, there will be less wave energy acting in the gap between the wheels over the quadrant where the paddles are down, causing them to be pushed together and the wheels to rotate. At closest approach the paddles are lifted to the horizontal position, so that the next quadrant of the revolution is not symmetric with the first; there will be no paddles in the water and nothing for the waves to act on. As they approach the first quadrant again, they dip back in. Thus we could extract energy from waves.

A similar nano device might possibly extract energy from the quantum froth of the vacuum. We have not really created anything, just borrowed some energy from the vacuum, a sort of perpetual borrowing machine, perhaps. Also, we don't know how the vacuum might take its energy back. We might for example create a hole in the vacuum. This might reduce the volume of space, for it is possible that space is kept 'inflated' by vacuum energy. This might be an acceptable trade off. However, there may be an outside chance we could invent a new type of black hole, one that swallows space instead of matter. Science fiction buffs could invent a new form of galactic travel: one could remove space in front of you in the direction you wanted to go, and dump it behind you to fill in the gap. An aeroplane propeller does more or less the same sort of thing, with air instead of space.

There may be a way of extracting energy from the vacuum by relative motion too, for example, if quantum particles form in a piece of space that is moving in relation to an observer. (If you don't like the idea of empty space moving, you can move the observer.) They are no longer symmetric pairs in the plane at right angles to the observer's direction of travel, for one will be going faster than the other at the time of division and therefore carry more energy in relation to the observer. There are a lot of questions we can ask of quantum particles. Unfortunately there are not as many answers.

If you happen to have a lab capable of asking this type of question, or do that kind of maths, here are a few for starters. Some have already been answered, but I will ask them to give the flavour of the thing.

1) Is it possible to define a rate of quantum particle formation? I.e., can we say that a given volume of space will always yield on average, the same number of events in a given time?

2) Is all space equal? Or is there a greater or lesser probability of quantum particles where there is a field? Does the proximity of matter or a gravitational or magnetic field enhance or suppress their formation? Does matter exclude particles, or can a quantum pair form inside an atom?

3) Is there a statistical spread of different energies? Is one more likely to have low energy events than high energy ones? Is there a cut-off point at either end of the spectrum?

4) Is there a dwell time? Does a high-energy event take a longer or lesser time than a low energy one?

5) Does the occurrence of an event inhibit the occurrence of a subsequent event, either in time or proximity? Is there a kind of 'loading time' as for a condenser charging, or a relaxation time? If so,

does a high-energy event suppress the occurrence of a subsequent one for a longer time than a low energy event?

6) If the observer is moving in relation to another observer, can each observer see each other's events? In other words, does a moving observer take his observed events along with him, or is there some kind of spatial framework in which events occur?

7) Is it possible to get quantum events to annihilate with the wrong partner? I.e., could one have a line of events where there was a negative particle left out at one end of the line and a positive one at the other, due to displacement of partners along the line?

8) Is it possible to get two events of different energies to interact? Can we get a single high energy negative particle to negate two positive ones of half energy, with reciprocal annihilation for the single high energy positive particle and the two lower energy negative ones?

9) Do events occur at quantum energies, by fixed steps, or is there a spread? Are these quantum steps always equal, or is there a scale according to the energy involved, with perhaps bigger jumps at higher energies? What kind of distribution can we expect; is it a Gaussian curve, like a bell curve, or is it skewed in some way, rather as the curve for gas molecule speeds is curved, due to the speed being a function of the energy level?

10) Do quantum events occur in space, or do they create the space they occupy?

11) Is there a theoretical anti particle pair, where a pair of negative energy holes is created in space?

12) Are there families of different quantum particles? If so, are these associated with different types of field? And is there a ratio between them according to field strength? Is there any interaction between events of a different nature?

13) How are events affected by the expansion of the universe? Do they get more dilute as the universe expands, or is the rate and density constant?

14) Very small particles exhibit Brownian motion; at a much smaller scale does 'Quantum Brownian motion' exist? Perhaps we could examine Brownian motion in supercooled helium with some

other atom as a marker, and see if there was a component of motion left due to quantum effects.

We can go on asking questions, however they might not all be the right questions. Unfortunately one can sometimes ask the right questions, and still not get any answers.

At a more mundane level we can examine negentropic devices. Maxwell made one such suggestion. He proposed a 'selective demon' which could observe the velocity of gas molecules and operate a trapdoor between two enclosed volumes of gas at the same temperature and pressure. The demon could open the trapdoor when it saw a fast molecule approaching, thus eventually trapping all the fast ones in one of the two volumes; similarly it could let all the slow ones into the other side. An ordinary heat engine could then be run between the two volumes until they were at the same temperature and pressure again. The engine would presumably drive a machine to make use of the work, eventually degrading its energy to heat, which could be recycled.

This is a good example of a perpetual motion machine, apart from the lack of a demon, that is.

So perhaps we ought to see if there are any demons about. I have heard that they are found in the detail so we will take a close look at random molecular motion. The best way to see it is by observing Brownian motion. Brown, a botanist, first recorded this when observing pollen grains under a microscope, and correctly inferred that unseen water molecules drove the constant jiggling and random motion he saw. Lucretius, a Roman writer describing Greek atomism, mentioned the possibility also, but recorded no observation. He also mentioned the possibility of electrons. I would like to have access to all his sources! When young I could just see a faint flickering in cocoa as a thin film on a silver spoon in bright sunlight. I can't see it now, so someone with good eyes could check this out for me.

I have seen disproofs of the demon to the effect that the energy required to sense a molecule would be greater than the energy gained, rather in the same way that one cannot increase the accuracy of the measured position of an electron without losing accuracy regarding velocity. This is in error, for we have a wide range of possible particles to measure, with different energies, whereas we can measure at a fixed energy level.

We can start with a demonstration. Suppose we draw out a thin glass tube, put in it a suitable particle and liquid, seal the ends and then balance it, in a vacuum, on the finest pivot we can come by. In its peregrinations the particle will sometimes occupy one end of the tube and at other times the other end, most time being spent in transit. It is possible that we might observe a tipping of our seesaw as the particle migrates. We might even be able to exert a magnetic couple at the ends of our see saw and drive an electron along a wire. A rather slow perpetual motion machine!

One objection might be that there would not be enough energy to drive an electron, to which we can reply that we can make the lever arm twice as long, or a hundred times as long if you like: a much slower perpetual motion machine. We could even put in two or more particles and wait even longer for all of them to finish up at one end together, which they surely will do, eventually. Another objection might be that, once the tube has tipped, the particle would have to go uphill to get to the other end. Well, that is just what Brownian particles can do. Half of the time actually, if you discount the time they spend going sideways.

A simple experiment can show this. If we set up a glass tube of water, and add some soluble coloured substance, copper sulphate for example, to the bottom, it will slowly dissolve and diffuse upwards. Over the course of days and weeks the blue colour will rise up the tube. You may even get a 'reflection' at the top where the copper sulphate molecules rebound off the surface and start to re-traverse the tube, leading to a darker band at the top. Copper sulphate molecules are a lot heavier than water, and work is done in getting them up there. The random thermal motion of the water molecules does this work, which will be slightly cooler in consequence, although this would be so slight as to be almost impossible to measure. Eventually the dispersal will be complete, although for a very long tube there would theoretically be a slightly higher concentration of solute at the bottom, due to gravity.

We can accentuate this density gradient by using larger Brownian particles. Suppose we could prepare iron dust fine enough to exhibit Brownian motion; we could put this into light oil as the dispersant, for it would oxidise in water. According to our particle size we would have a model atmosphere of iron particles, with a density distribution down the tube, just as air is denser at the bottom in a real atmosphere. If we wished, we could place a strong permanent magnet at the top. Eventually most of the particles would traverse the tube and be trapped by the magnet. We could thus show that the magnetic field had acted as a Maxwellian selective demon, trapping all those particles with sufficient energy to climb the tube, and not letting them go back again.

It might seem at first as though the energy required to pull the particles off the magnet might equal or exceed the work done against gravity. However, there is a difference between the gradient of the two fields. Gravity can be considered to have its focus at the centre of the Earth, four thousand miles below. The few metres or whatever length of the experiment will hardly make much difference to the ratio of field from top to bottom; we can consider it as a uniform field. This is not the case for the magnetic field however; this will also obey the inverse square law, but from a much more immediate start point. Thus we could have a pivoted tube, which can be turned upside down, putting the particles back at the bottom again. Initially, energy will be required to pull the tube from the magnet, but then the excess weight of the particles will take over and the heavy end will fall, liberating gravitational energy. We can make the tube as long as we like, so we can arrange for more gravitational energy released than that amount required removing the particles from the immediate vicinity of the magnet. The excess energy is provided by random thermal motion. Another very slow perpetual motion machine.

We can even do it with a solute in water. Seawater contains a few per cent of salt by weight, and this exerts an osmotic pressure. This is best shown across a semipermeable membrane, which can be thought of as acting in a similar way to a sieve; water can pass either way through the membrane, but salt can't. Thus there is less water going from a salty solution on one side to pure water on the other, for some of the pores are blocked by salt in one direction only. The pressure developed is finite, although it can be quite large. If one pressurises the salty side, one can eventually force the water back against the natural flow, for now the water molecules on the salty side have more energy than those on the other. This is used in some desalination plants to make fresh water from seawater; it requires less energy than either evaporation or freezing and works even better in brackish water, for the lower the salt concentration the less the pressure required. There is a plant in Malta, opposite Gozo, should you ever go there. Fifty years ago the groundwater was already getting salty, due to over extraction by lots of little aero motor windmills, for irrigation. We used to get our water from the dockyard still when I was there. Most of the aero motors have gone now, and most of the goats too, and the water is quite good. Personally I would have got a few old single bottomed oil tankers, capable of carrying a quarter of a million tons or so. They are very cheap, only a few tens of pounds per ton as a scrap value. They could then be filled up in the estuary of a big river, like the Nile perhaps, or a nearer one in Italy. Running at eight knots or so the energy cost of transporting the water would be less than desalination. You could even combine the two; river water could be pumped back into an aquifer to act as a filter and reduce the cost of reverse osmosis, rather than using seawater.

If we take a very long tube, with one end sealed by such a membrane, and lower it into salt water, at a few thousand feet fresh water will be squeezed out of the salt by the pressure of the water above. Fresh water is less dense than salt water, so the column of fresh water climbs higher and higher up the tube the deeper you lower it. At something over twenty four thousand feet the fresh water column breaks surface, and at deeper levels you have a head of fresh water that can be used to drive a toy turbine. You might need rather a deep ocean trench, but the Marianas would do, and you might need rather a large surface area of membrane and still have a rather slow perpetual motion machine, but the principle is there.

Given that such a device will produce energy, we can ask from where this energy comes. It is of course provided by gravity, with the average centre of gravity of salt in the ocean being lowered. Theoretically, the salt will slowly diffuse up again giving a continuous cycle, and the random thermal heat of the ocean is converted into work. A marine demon!

In practice of course the ocean is stirred by wind, heated by the sun and undersea volcanoes, and deflected by coreolis forces as it moves, so we would have difficulty in actually demonstrating our effect independent of these much greater variables, but it should be there nevertheless.

We can speed things up a bit by using air instead of water; we already have an atmosphere with a good pressure and temperature gradient. At the surface air pressure is about fourteen point seven pounds per square inch; at eighteen thousand feet it is half that. Similarly, for every three hundred feet you climb, the temperature drops about a degree Fahrenheit.

So why does it do this? Again, it is gravity, which is beginning to look increasingly like a Maxwell's demon. It can discriminate absolutely between a fast and a slow atom, with no energy cost in doing so. In a standard model for a gas in a container, we explain the pressure as being the result of billions of impacts made by the atoms or molecules of the gas with the walls of the container. Increase the temperature, and they go faster and hit harder and more often. Increase the pressure by putting more gas in, and there are more impacts because there are more atoms to cause them. This is all very correct and self-evident. We can extend the model by adding rotational energy, and oscillatory energy for the individual atoms in a molecule, but this is just detail. Most of the time, unless the gas is very dense, the atoms proceed along a 'mean free path', which is that average distance during which they proceed in straight lines between impacts with each other.

Actually, a gas is mostly empty space, so most of the time atoms are unperturbed by each other, giving an average velocity for an atom. For hydrogen at room temperature it is well over a thousand miles per hour, so there is quite a bit of energy there. This is why we don't have hydrogen in the atmosphere, while the sun does. Sufficient hydrogen atoms in the earth's atmosphere will attain escape velocity and evaporate into space. The sun has a much greater gravitational field, over a longer distance, so, although hotter, hydrogen cannot escape from it to any degree. The other gases in the earth's atmosphere are too heavy and their chances of escaping are far lower. There might even be a trace of xenon left on the moon, which has such a low escape velocity that a nineteen forties V2 rocket could just about struggle off, if you could get it there in the first place, that is. Somebody raised the myth that there was a B47 bomber on the moon; actually it is a No 47a London bus, but don't tell anyone! Having said that, the mean free path is rather short: atoms collide several million times per second, which is just as well, for otherwise we might have a bumpy ride, as some microscopic water creatures do. Presumably they get used to being bombarded by (on our scale) things the size of peas, as it is just part of their environment.

At the large scale however, the concept of constant velocity straight-line free paths breaks down. The paths are not straight, unless they are going straight up or straight down, neither are they of constant velocity, unless you count those that begin and end on the level, adjusting slightly for the ballistic curve they trace. Neither are they totally free, for gravity affects them all. At first glance it might seem that, for an atom travelling a hundred millionth of a centimetre, for a few trillionths of a second, this effect would be totally negligible. After all, what is calculus for, if not to get rid of these inconsequential fractions? All you have to do is to say dy, dx and they go away!

Unfortunately for this argument, all these events are of the same sign, and they are cumulative. There are billions and billions of very small events per second in a cubic centimetre of gas, and they are all deflected in the same direction. All paths not vertical are curved downward, and all speeds are either reduced if they are going up, or increased if they are going down.

The net result is what you see. The air is denser at the bottom of a mountain than it is at the top, and it is also cooler. You may have seen the statement 'air expands as it rises, and cools'. This might all seem very logical, but there is an inherent assumption embraced within it that is false. Most people seem to assume that the air cools because it expands. Actually, expansion of a gas does not cool it. The velocity, and hence the temperature, of a gas molecule is exactly the same before and after expansion. It will only cool if it does work. For our atmosphere work is done against gravity. If one pound of air rises fifteen thousand feet, then fifteen thousand foot pounds of work have to be done. This is equivalent to about five hundred miles per hour loss of velocity for our molecules, or about half a horsepower minute per pound weight transported.

So can we use this effect? We could for example take a tall mountain; Everest springs to mind, but it is rather difficult of access, and part of a chain. Kilimanjaro might be better, as it is over nineteen thousand feet and starts from a plain at a mere three thousand feet. We could run a well-insulated pipe from top to bottom, set up an air compressor at the top to compress air isothermally, run the dense, cold air down the pipe and expand it isothermally, via a heat exchanger through an air motor at the bottom.

There should be a net gain of energy, with the air motor developing more power than required by the compressor at the top. We could even liquefy the air and have aero-electric power instead of hydroelectric as well. If you don't think it would work, how do you explain hydroelectric power? It is exactly the same effect, running to a natural cycle. There is nothing more random than the movement of evaporated water molecules in the atmosphere. Yet, when they condense at altitude, we suddenly have a macroscopic event, where a mass of water can take a non- random return path back down the gravitational field. Somebody has even proposed building a very tall structure, or one held up by a stratospheric balloon, many miles tall and using hydrogen as the working fluid. If we ever get a space tether, attached to a mass in a geostationary orbit, it would be an attractive proposition. The main elements of such a plant would be:

1) A gas which is easily liquefied at the temperature pertaining at your chosen height, the height itself being dependant on latitude.

2) An insulated pipe down several thousand or tens of thousands of feet to warmer air.

3) An expansion motor, using ambient temperature to extract energy from the gas as the liquid boils.

4) A much larger pipe back up to altitude, to feed an isothermal compressor to liquify the gas if it does not do so by itself. Not strictly necessary, as it will find its own way back.

5) A 'carrier gas' may be used in the same way that air carries water vapour up to be condensed if this seems a suitable route to take.

Purists might argue that the sun powers the whole effort, and it is not perpetual motion. This is correct: the planetary air circulation and the water cycle absorb several thousand horsepower per acre each day, and we have already shown that there is no such thing as perpetual motion due to a lack of time anyway. However, if you cannot succeed, cheat! The main game is a possible power source, not proving the unprovable. We don't have to wind the whole Universe back up, just a little bit of it; we can use a slight squeeze instead of a big crunch. For those who still doubt, don't worry; all we are actually doing is exploiting a temperature differential, which is what all heat engines do.

Will such sources ever be developed? Probably not yet, other than as a proof of concept device. The plant would be large and expensive per horsepower hour compared to other devices. The temperature differential is not large, air is very bulky stuff and the cost of insulated pipes up mountains would be a downside. We might do better to use water to carry the heat; it would be less bulky and the warmer water would rise by itself, obviating a pumping load. Then of course we might find that most suitable mountains are not near centres of population. We could of course use smaller mountains; five thousand feet would still provide a useful temperature and pressure differential, and we could increase the efficiency by increasing the working pressure. We could even build a mountain, or rather a mast. Transmission masts can be over fifteen hundred feet and built out of the skimpiest amount of steel. The new carbon nanotube materials could be used to go miles rather than feet. Probably the best way however would be to produce our own gravitational field by means of a centrifuge, for we can easily make a few hundred thousand 'g' in a short compass.

It is difficult to predict the future of energy production. Also, demand predictions may be well off the mark; currently increased efficiency seems a very viable route, as demand might actually drop in some countries, with slower growth in others. It might however be instructive to build small-scale models to demonstrate the principle, for should such sources be used, it is certain that they will be more sophisticated than the bare outlines here presented, and there will be a development trail to be followed.

At least we have shown that it is theoretically possible to extract energy from random thermal motion, using a gravity field to perform the office of demon. As such we have a model for a second order thermodynamic perpetual motion machine that could be made to work. We have shown that 'chance favours the trained observer'. However, drawing on long experience, I can accurately predict that the general response to perpetual motion devices will be 'chance would be a fine thing!'

A short word picture here, as a light break from reading a long article.

When I was young, I used to go camping around Europe on my 200 cc Triumph Tiger Cub, top speed 65, cruising speed, 65, and 125 mpg two up, and half a hundred weight of gear.

Once, in Holland, I fell into conversation with a Dutchman, and enthused about the Dutch Daf car, which had a gearbox composed of two opposed tapered cones, with belt drive in between, an idea later pioneered in a British scooter and now in all of them.

"I won't buy a Daf," he said.

"Why not?" I enquired.

"They collaborated with the Germans during the War!"

"So what do you drive?" I asked.

"A Mercedes!" he replied.

SOME EXPERIMENTS WITH GYROS

Gyroscopes make fascinating if puzzling toys; although no more than spinning wheels, many of their characteristics appear counter-intuitive at first sight.

To begin with, I first wish to talk about spin, as many of the problems people experience in thinking about it are due to a lack of a mental model of what spin is, and how spins in different planes can or cannot be combined.

Surprisingly, there is very little written about spin. Perhaps it is so simple that nobody has studied it in detail; perhaps there is not much to study, although it is one of the most fundamental characteristics of the universe. Galaxies spin, or most of them do. Suns spin, planets spin, electrons spin, both individually and around the nucleus. Even light has a spin component, and nuclear physicists talk of 'spin' for elementary particles, although they might not mean quite what you and I mean.

Newton wanted to know why a spinning bucket of water heaped up round the edges. Did it in some way 'sense' the rest of the universe? Once, when at school, I tipped all the lab's mercury into a ten-inch pneumatic trough, and put it on a gramophone turntable, with the idea of making a big parabolic mirror for a telescope. I was rewarded by seeing a large eye peering back at me, but there were lots of cross ripples, probably because the turntable was overloaded. Nowadays that is the way they make them. Does the universe spin? And is this a meaningful question, in the absence of any reference point apart from the universe? We could construct a model in which the proto universe spins in three planes, thus giving expansion and possibly explaining why the expansion seems to be increasing incrementally as we go outwards.

189

The early astronomers had difficulty with it. A few hundred years ago it was assumed that the sun went round the earth, although the Egyptians and their predecessors knew better, but this led to some odd paths for the planets, the inner ones reversing their paths to form loops or epicycles, while the outer ones varied their speed. Suggesting that the earth and all other planets rotated round the sun ironed out these problems, although it took time for others to agree.

Although it seems simple now, given only an observer and something for him to observe, it is not that easy to determine who or what is spinning. Without any other reference, it is not possible to say whether an observed object is rotating around you, or if you are rotating and the object is fixed, or if both are rotating at once. Once we bring in other objects to observe, and maybe centrifugal or coreolis forces, the problem becomes solvable, although even then there may be alternative explanations to be discarded

Compound spin, where there is spin in more than one plane, is difficult to sort out conceptually. I will give some examples to show where the difficulties may lie. We can do this by means of a series of Gedanken experiments. To start the ball rolling, so to speak, I would like you to imagine a large ball, like a beach ball, so we have something to do experiments with.

First of all, put a blob of paint on top of it to show where the North Pole will be when we rotate it. You can put a dot of a different colour on the equator, to show when it is rotating. If you now spin your beach ball equatorially, you will find that the dot at the top remains stationary, while the one on the equator spins round, tracing a line which bisects the ball into a lower and upper half, or Southern and Northern hemispheres.

Next, I want you to stop the ball spinning, and fly round it yourself, equatorially, to see if there is any difference. You should find that the observed results are identical. There is no difference between a ball going round, and you going round a ball, observationally at least. If you can see the background, or have a little weight and a spring balance, you can soon tell the difference, but by mere observation, there is nothing to tell the two cases apart.

Now I wish to combine two components of spin. To do this we can attach two little rockets to the ball, one firing along the equator, so as to give an equatorial spin as before, but the other at right angles, so that it would, if it were the only force, spin it pole to pole. It might be an idea to float the ball in water, to stop it rolling around; for this experiment, both rockets are fired at once, giving equal thrusts. The ball cannot move in two directions at once, being a solid object, so what you should find is that a resultant direction of spin, at forty-five degrees between the two rockets, ensues. Your North Pole will now trace a circle, and the ball will establish a new pole, so that it is tipped over at forty-five degrees, rather as the earth is tipped over at an angle to the plane of rotation about the sun, except that we are tipped at about twenty-three degrees.

So far, so good. I expect you can guess what comes next! For your next Gedanken experiment I want you to rotate the ball equatorially as in the first case, and then fly round it in a little aeroplane, at right angles to the direction of spin (that is, over the poles) at the same speed as the rotation

You can do this stage by stage, doing a quarter of a turn at a time, if you find it difficult to see what happens. You should find that the two spins do not now combine into a resultant as before, but rather you seem to trace a figure of eight path over the surface of the ball, and you will only ever see one side of it. To many, this is a surprising revelation. It is surprising because it is difficult to think of two spins at once.

I said that it is impossible for a solid object to spin in two different directions at once, and this is rather self-evident. However, there is another experiment you can perform that seems to contradict this.

Float the ball in water as before, and give it a spin equatorially. Then get a hosepipe and direct a jet impinging downwards at the equator. Although it looks as though you are applying a force at the equator, which should give a resultant as with the two rockets, you are not doing quite the same thing. The force applied from the hose pipe is fixed in one position, while the ball turns beneath it. You should find that your North Pole starts to describe a circle, which expands as the ball tips more and more. It then passes the equator and contracts to the position previously occupied by the South Pole. The ball is still spinning as before, but the poles are reversed. This is sometimes called a 'Tippe Top' reversal, not because it tips, but because the inventor was called Tippe. Sometimes you can find a flat ovoid stone on the beach, which if spun flat side down will suddenly

191

stand up, so that it is spinning with the long axis vertical. I will leave you to work out why this is so. You may be able to buy a Tippe Top from a supplier if you can find one. It is not much different from an ordinary pear shaped top, but if spun heavy end down, it will quite suddenly tip up so that the heavy end is on top.

Thus I hope to show that spin is not quite as simple as it first seems, and we should never take anything for granted. If you are very good at visualisation, or perhaps maths, you may be able to derive more complex experiments, combining, or failing to combine, spins in three planes, or even hypothetical further dimensions. If so, I hope you publish your results. If you find that your maths does not correspond to the results of experiments, then your maths is wrong!

Now a gyroscope, if set spinning in a vertical plane, and supported at both ends horizontally, spins like any other wheel. However, if we remove one support, so that it the centre of mass is now one side of the support, it does not fall down as it would if not spinning, but starts to precess. That is, it starts to rotate about the support, in a plane at right angles to the spin axis, with the top of the gyro progressing retrograde to the movement around the support.

Now, why on earth should it do this?

Obviously, we are dealing with compound spin in two planes, and the force of gravity acting on the gyro. In some ways it exerts a torque, or turning force, so that all the weight is transferred to the supported end. It is a mistake to assume that the other, free, end is held up by thin air. If this were so, then the weight at the support would only be half, as in the case when it was supported at both ends. Simple recourse to a spring balance will show that this is not the case: all of the weight is transferred. This is quite interesting, as we have managed to separate the centre of gravity, where the weight acts vertically to the support, from the centre of mass, which is obviously precessing round this point. It might be possible to step out, i.e. sequentially moving the mass sideways by a series of half precessions, with no opposite reaction, but I do not know if this is possible. Experimenters please note!

To show that it is an effect caused by gravity, it is more convenient to tie a string to one end of the gyro, and suspend it from a spring balance. If one then lowers the gyro while it is accelerating downwards, the precession slows, and it weighs less. If you accelerate the gyro upwards, the precession increases, and it will weigh more. Further, if you hang a small weight on the end that is precessing, it will precess faster, to compensate for the extra load. If you increase the spin rate, the precession rate slows. If you let go of the string and allow it to fall freely, precession stops instantly. The fact that the string hangs vertically shows that there is no centrifugal effect caused by the precession, which is of interest. Also, there is no energy in the precession. Interestingly, when the gyro strikes the floor, the centre of mass is to one side of the point where the string suspended the weight of the gyro. Repeat the experiment, allowing the gyro to go one half precession further, and you can have the impact point on the other side of the central point. It seems that you are able to achieve a separation between the centre of mass, and the centre of gravity.

What happens is that the circular spin path opens out to an epicyclical path, and the spin rate slows, to compensate for the greater distance embraced by this path. Letting go of the string causes the epicyclical path to immediately close up to a circular one again. If you force the precession, that is, give it a little push to try to make it precess faster, it will tip up, so that the precessing end is higher than the supported end; if you retard it, it will tip down. I am not sure if the gyro weighs more while it is in the process of tipping up, or whether your push is turned through a right angle, with no increase in torque. I think there is no reaction, but others say there is. So can we make a flying machine by causing a precessing gyro to rise? Up to a point, but you will only get a reduction in weight while it is accelerating upwards. This cannot continue for long before your spin rate exceeds the tensile strength of the material. As gravity is providing a constant force accelerating downwards, you will not get very far.

However, you might be able to make a device for moving things around the outside of a space station. This would consist of a pair of gyros instead of one, fixed to a couple of flexible joints like your wrist and shoulder. For the reaction mass, you would need another pair, of opposite sign; that is, spinning in the opposite direction and being forced to precess in the opposite direction so that both pairs move in the same direction. The speed would be a function of the spin speed and would be limited by the strength of the material. A space wheelbarrow rather than an interplanetary device, but useful all the same.

A word of caution about torque here; you may have used a torque wrench, which is a kind of spanner used for tightening nuts, but not over tightening them. A ratchet slips when the correct force is applied, and it can be set for different degrees of tightness. When using such a wrench, you are not applying a true torque, as will be evident if you try to tighten the front wheel of your bicycle by heaving upwards. The front wheel will come off the ground. You are applying a torque plus a lift. A true torque will be developed by one of those wheelbraces in the form of a cross, where you pull up on one side, and push down on the other.

There is a problem with a gyro here. If you take a heavy gyro, non-spinning, and try to hold it out at arm's length, it is very difficult to do so for long. However, once set spinning and precessing it is easy. If you try to force the precession, to make it precess faster, it will rise and be very difficult to hold down. Our simple explanation, that the gyro exerts a torque, will not serve here. First of all, it is exerting a torque through several flexible joints; the wrist is a compound joint, which can move in any direction, the elbow is a hinge, and the top of the arm has a ball and socket joint, again capable of moving in any direction. This is a rather funny kind of torque, somewhat equivalent to using a piece of string as a crowbar! Also, the gyro, if allowed to do so, will not tip, but rise rotating in the same plane. It rather appears that the force exerted to try to make it precess faster, has been turned through a right angle to make it rise, with no obvious reaction. If you try to mimic this action with a non-spinning gyro, you will have to apply a vertically acting force to get it to rise, and it will appear to weigh more whilst being accelerated upwards and less while it is decelerating, before stopping in its new position.

So there are some real experiments to be done here. A characteristic of a gyro is that it responds at right angles to an imposed force, as anyone who has tried to twist the axis of a gyro while holding it will know.

So far we have described what happens, but with no explanation as to why it happens. To see if we can get an explanation we can examine what happens at one 'mass point' on the gyro, as this simplifies things, although of course we know that the gyro is really composed of a large number of mass points. We will also have a 'perfect gyro'; that is, one where all of the mass is at the rim. In a real one some of the mass is obviously in the axle and in the metal supporting the rim.

To do this, we can imagine a small spring balance, with a small weight, say one gram, suspended from it, at the rim of the gyro, which is in the horizontal plane, like a top. Before we spin up the gyro, the weight hangs vertically, and the balance reads one gram. As we start to spin up, the weight begins to hang out at an angle, like a chair on a chair-o-plane at a fairground. The reading goes up also, for we are now reading a compound force made by the original one 'g' acceleration of gravity, plus the centrifugal force, at a right angle, engendered by the change of direction of the mass as it performs it circular path. I use the term 'centrifugal' as it is in common usage; the force is really centripetal.

Very high-speed gyros, used for separating different molecules in solution, can pull up to a million 'g', I am told. They run in a vacuum, behind several inches of armour plate too, as their energy can be equivalent to over their own weight in explosive, so please don't be tempted by high speeds, and don't try to break the world record; it stands at ninety million revs per minute! (Footnote, page. 197, *Propulsion Without Wheels*, Prof. Laithwaite, ISBN 0 340 04836 0.) All the experiments you need to do can be done at not many revs per second!

Professor Laithwaite did a lot of experiments with gyros, but died unexpectedly before he had finished his work. He used to phone me occasionally from his lab in Sussex; I do not think his work was carried on. Some people criticised him for pursuing unrealisable goals. As I read it, he realised that he did not know all there is to be known about gyros, and wanted to find out, whereas his critics knew all that was known and thought that was the end of the story. It is much more difficult to extend a knowledge base than discover what other people have done. In fact too complete a knowledge can have an inhibiting effect. There is hope for us yet!

I had a copy of the patent spec for the Dean Drive once; this was a device that purported to move along with no Newtonian reaction. It consisted of a pair of eccentric cams, which were released to fly away for a short distance; then the rest of the device was cranked up afterwards, a kind of 'bootstrap' effect. If you built one, I suspect it would merely wobble. There was another inventor, In the Channel

Islands, who had a device that purported to do the same thing. This consisted of a gyro, mounted at 45 degrees, which was forced to precess in a cone. A chap from the patent office phoned me about it once; he said it had a great propensity to move downhill! I suggested he mount it in a box, for a gyro thus moving can act as a sort of inefficient aeroplane propeller, and float it on water so there was no downhill. I don't think he did though.

I used to write a bit of science fiction, and a friend tried to get me and a mathematician to work on a non-reactive device. Unfortunately I had also read a science fiction short story called 'Noise Level' in which a group of scientist were shown a short film of a levitating device, purported to have been captured from the enemy. Although known as an impossible problem, they beavered away and built one. We did not build one but came across some interesting bits and pieces in the sidelines. Perhaps if I had not read the story we would have succeeded!

Sperry did a lot of work on gyros, some of which is still classified. Nowadays they use 'light gyros', which are not gyros at all but triangular glass blocks with a light being refracted round them; rotation alters the wavelength which can be measured by interferometer.

Here is another device which forms a good classroom experiment and a test for students:

You need two opposed equal pendulums; a couple of bars of metal with weights at the end will do, and a flat box to put them in. When in the 'up' position they may be locked by a small weight placed between the two short ends, which do not protrude into sight at the top of the box. Having balanced your box with a counterweight over a couple of pulleys, you ask the students:

'What will happen if I remove the weight?'

They of course say the box will rise, as it is no longer counterpoised, unless they have already sussed you as a sneaky kind of lecturer. You remove the weight and the box goes down, in a series of starts and stops, the reason being that as the pendulums fall, they generate a centrifugal effect.

You can also have an upside down pendulum, by using a magnet against an upside down iron retort stand base. You can ask your students, 'Does a ticking metronome weigh less?

Another example; you are the cashier at a theatre, and crooks have held you up. To keep you out of the way, they use some stage props, consisting of a gilded cage, on a counterpoise, and leave you suspended twenty feet above the floor. How do you get down?

The answer is simple; all you have to do is to pump up the cage in the same way as a child may get a swing moving. It is more or less instinctive, as we spent a lot of time in trees. I have seen monkeys do the same thing to get a branch moving before a jump; the centrifugal effect plus the weight move the cage downwards. If you watch a child on a rope swing over a branch you will see the branch goes up and down for the same reason.

Yet another example: it is easy to find the centre of gravity of a sheet of card, all you have to do is to stick a pin through it, hang it up, and draw a vertical line from the point of suspension. Draw another line from another point on the card and you will get two lines that cross at the centre of gravity.

If we now tip the gyro so the axis is horizontal, supported at both ends and not precessing, we find that the two forces of gravity and centrifugal force now interact cyclically on our spring balance; at the top of the wheel, gravity is acting downwards, as always, while the centrifugal force is acting upwards. When at the bottom, they act together, with a simple harmonic variation in between. If we plot the reading on the spring balance, we will get a cyclic curve as the weight regularly goes from centrifugal force plus gravity, to minus gravity.

Next, we take one support away, or more simply hang the gyro on a piece of string with the axle horizontal, and watch it precess. If we trace the path of our mass point we will see that it follows a series of looping curves. To get some idea of this path, imagine that you are standing next to a white wall, with a piece of crayon in your hand, and you draw a circle on the wall, perhaps some three feet in diameter. Keep on drawing, and you will soon get a more or less perfect circle. Next, get a friend to push you along the wall on a skate board, drawing your circle all the while. You draw so that the top of the circle is retrograde to your movement along the wall. You will find that you have drawn the kind of path traced by the mass point on a gyro. As your crayon is going more slowly at the top, the loop formed is tighter there, compared to the bottom, where your rotational speed and the speed of the skateboard add together.

One further step is needed, for this series of epicycles is linear, and we want them in a circle. Just rolling up the wall into a cylinder will not do, for it will only curve from side to side, and not top to bottom. What we need is a room that has curved walls that conform to the inside section of a torus, so that you can draw your circles at a constant radius. A small roundabout such as can be found at a children's playground completes our equipment and you can now exactly trace the path of a mass point on a precessing gyro! You will notice that it is actually a curve in three dimensions, not two as originally supposed!

We have not quite finished yet, however The next step is to compound all of the vectors; that is, all of the directions our spring balance points in, dangling its weight, and the readings that weight produces. This will show the net force acting on a series of mass points around the gyro as it precesses. Some computer help would be most welcome here. This can perhaps be more easily understood if we ask the question, 'what would happen if?'

We may wonder why the gyro does not just fall down, as it is unsupported at one end. Well, if it started to, then the speed of one side would increase, as we would have the spin speed plus the acceleration of gravity, while the other would slow down. Similarly the top would have to fall outwards and the bottom inwards; again, this would alter the speed and the centrifugal force engendered.

In one sense, the gyro is falling, but it compensates by moving round in a circle. You can think of this as a species of orbit, if you like. Newton asked the question, 'Why does the moon not fall down?' and arrived at the conclusion that it was in fact falling, but as it was also going round the earth, it would always keep at the same distance above the world. If you fire a shell from a gun, it will cover a certain distance before it meets the ground. Double the speed and it should go twice as far; you will need four times the powder though, as your shell will need four times the energy. Guns don't scale very well.

For very long distances it will go more than twice as far, for the horizon is dropping away under it as it goes round the curve of the earth. Get the shell to go fast enough (about eighteen thousand miles per hour for a low orbit), and it will never come down, air resistance excepted. I once had a great argument at college, with some maths students who maintained that the path of a shell is a parabola. I pointed out that, unless it achieved escape velocity, it would have to be a section of an ellipse, unless the earth is flat that is. They did not want to project the path though, because the ground got in the way!

Thus, whether we can work it out or not, the net result of the forces acting is that the gyro exerts a sort of torque, and rotates around its point of support, with no net centrifugal effect exerted at that point, otherwise the bottom of the string would describe a circle.

You will notice that I have not used any maths here. There are two reasons for this. One is that I am not very good at maths, and some of the readers may not be either. Also, just knowing the maths may mean that you just apply the formulae, and don't understand what you are doing. Thus it is sometimes better to go the long way round. Maths is good stuff though; I look forward to somebody extrapolating in 'n' dimensions, as this is not at all easy conceptually.

A good bought gyro will have three guard rings, to prevent you getting your fingers mixed up with it. Don't buy a cheap toy one; they are usually unbalanced with bad bearings. I make mine with old bike wheels, as they are well balanced and have good bearings. Lead pipe can be substituted for the tyre to give more mass, but don't go in for high speeds, and make sure it is well secured.

One frustrating thing about a gyro is that you cannot easily apply forces to the moving rim. However it would be possible if one could make a liquid gyro. For example, mercury is very dense. If one could form a pipe into a circle, and pump mercury round it (it will pump electromagnetically) one would be able to pick the gyro up by its rim, and do a whole new series of experiments applying forces there. Further, it need not be round! You could have a square one if you like, or a triangular one. More interesting would be an elliptical one. It might have two different rates of precession, according to which axis you had upwards. You could even have one with two different diameter pipes; a thin pipe round one half and a wider bore of twice the section round the other half. The centrifugal effect varies as to the square of the speed of flow, so you might be able to invent one that developed more centrifugal effect on one side than the other. More likely it would establish a new centre of rotation or precession however, but you might be able to get some striking effects for lecture demonstrations. Don't be tempted to think that you can make a gyro that moves sideways though. While it is true that the half bore section will exert twice the centrifugal effect of the full-bore section (not four times as much, as there is twice the mass of liquid in the full bore section), there will be a reaction force as the mercury accelerates from the full bore section to the half bore, and an equal opposite reaction when the reverse occurs. Unfortunately they are both of the same sign, as the direction has reversed 180 degrees as the liquid goes through half a circle. The net result is that these two forces cancel the extra centrifugal effect. However, what is true for reactions in the plane of the gyro need not be true for reactions in other planes. What is really needed is a good computer model of a gyro, where we can do all kinds of experiments from a desktop, before bending metal and pumping liquids. Computer buffs please note.

There are a few problems with liquid gyros however, including the possibility of turbulence. I will use the moon as an example to show what I mean. One might assume that the moon does not revolve about its axis, as it always has the same side to the earth. However, a moment's reflection will show that it in fact does rotate once about its axis for every rotation about the earth. Similarly, if we for example used a train of ball bearings in our 'liquid' gyro, they would rotate once about their axis with every precession; in a liquid this might lead to turbulence and rob it of energy. This is why a centrifuge, as used to train astronauts, can be sick-making; the little otoliths in your ear know they are rotating once about their axis in every revolution. We could however use a linked chain of ball bearings, similar to that type of bath chain that consists of a string of balls with rods in between. Thus we could prevent this kind of rotation. A mathematician might like to work out what extra forces would be generated in a liquid.

More interesting still, we now know the precessional path followed by a mass point around a precessing gyro, and we also know the net result of forces engendered by such a mass point following such a path, which is more than most people, I suppose. We can use this knowledge to good effect, and wind a pipe to mimic this cycloidal path, and pump a liquid or ball bearing train through it at the correct speed. We can call this a 'pseudo precessional gyro', and if you make one it will be the first one in the world. I don't have a workshop, and am not very good with my hands, so I have not made one. Anyone who has handled a fire-hose will know that liquids moving in pipes can exert some surprising effects though. Your new gyro will of course just sit there and do nothing until you try to move it of course. Then you will unbalance the forces quietly sitting within it. I will leave you to work out what possible effect you might get. Twist it and it might rise in the air. So that is how flying saucers work!

Inside the rim is a cycloidal bird's nest of pipes, probably filled with pumped mercury.

If you decide to make one with mercury, it might be an idea to warm your mercury to about thirty degrees, and then slowly stir in some lead shot. Not bird shot, that is sometimes hardened with arsenic; the kind used by fishermen would do, or make some lead filings with a rasp. The lead will dissolve in the mercury, and the melting point will rise, the advantage being that when you spill your mercury, it will set on the floor and be much easier to collect up.

Having done all the previous work, we are entitled to speculate; a kind of 'silly season'. We can propose all kinds of silly ideas, and perhaps some of them might be testable. Don't be inhibited in case you are wrong. There is nothing wrong with being wrong! What is wrong is not to speculate at all.

We will do another Gedanken experiment, this time in a boat; a long thin punt will do. First of all, you start at the back of the punt, which can be conveniently alongside in an inside swimming pool, to cut any wind effect. You will find that as you walk forward the punt goes backwards, in inverse ratio to the two masses. The centre of gravity of your weight plus that of the punt moves not at all however. Next, you put a turntable with a heavy gyro in the middle of the punt. Start with the gyro not spinning, supported at the centre on a pivot, and the other end on a roller-skate. Now rotate the gyro along what would be its precessionary path, were it spinning. You will find (apart from side to side rocking, which you can compensate for with outriggers.) that as the weight goes to the rear of the punt, the punt goes forward. As it completes the next half revolution, the reverse occurs.

You have exactly the same situation as when you moved back and forth in the punt. Next, start the gyro spinning and remove the roller skate. This time the mass will perform the same path as before, but the weight will only be applied at the central pivot. There is no net centrifugal effect about the precessionary axis, and the punt will not go back and forth. If you check the gyro at one half revolution, putting another support under the free end to stop precession, then slide it back to its start position, you can remove the first support and let it precess again, in the other direction this time, as the other end is supported. You should find that as you slide the gyro back, not precessing, there is the usual Newtonian reaction, but not as it precesses forward. You could repeat the process until you have moved the punt entirely outside its original position. Or maybe not. You could give it a go and find out though! I suspect that there may some reaction as you check the gyro in its precession, but I have been unable to find it.

One reality check would be to see if there are any other examples of non-Newtonian reactions. Surprisingly, there are one or two. For example, light has no mass, but it exerts a reaction force. Shine your torch, and it pushes back very slightly in your hand. This is not measurable, but if light impinges on a very small dust particle, it will push it along, unless it is a graphite particle, that is, which will spiral towards the light, due possibly to an induced current reacting with the spin of the photon (I'm guessing!). Solar 'sailing ships' have been suggested, where a space probe unfurls a big sail and is blown away from the sun by light pressure. This is not strictly Newtonian, for we have a zero mass item exerting a reaction on a mass. Similarly, it is possible to rotate a plate by an induced current, causing a magnetic effect. The 'Newtonian' reaction however is not against the rest of the apparatus, but against the magnetic field.

We know that we can turn the force of gravity through a right angle, so that it appears to travel along the central spindle of the gyro, then through another right angle to reappear as a vertical force once more, with the weight seemingly all displaced to the point of support. Suppose, by some more complex mechanism, we could support that end also, we might get a new kind of hovercraft, a sort of orbit that works in a small compass, at a lower speed more suitable to the small section of the gravitational field it covers.

I once read an old Indian manuscript that described 'vimanyas', or flying machines. It was translated in Victorian times, before flying was invented, so they could not make head or tail of it. Nowadays it is regarded as early Indian science fiction about air warfare, or a myth. One thing that stuck in my mind though was that it was described as running on mercury, which is hardly the stuff you would want in a flying machine, unless you had a very good reason. Early peoples moved very large weights; there is a canyon in South America with a six hundred ton section of a carved temple in it. The quarry is known, and the temple site is known. The point is, how on earth did it end up at the bottom of a canyon? Well, they dropped it whilst moving it between the two, obviously. But how were they moving it? A rope bridge would have to have had some very substantial ropes, and the anchor points would still be evident. If you are out that way you could have a little dig underneath to see if you can find any mercury!

It is good fun to speculate, but perhaps it might be best to do the experiments!

ON PLANTS, BUGS AND OTHER

ITEMS

Sometimes one comes across odd pieces of information that may be of use to others, or may be of general interest. I expect most people have had the experience, but do not know what to do with it other than perhaps tell a friend.

I think quite a lot of good scientific observations may be lost in this way, so I thought I would set a few down. They are just 'fillers', of no great consequence in themselves, but they may be of use to someone else doing work on the subject. A publication needs fillers, not just to fill in the bit left over at the end of a page, but to make short bits of reading to break the thing up. I hope other people will do likewise; it does not take much time! The sum total of such observations could be of great use, and in a very small way one may contribute to 'real' science.

We had Dutch elm disease in Britain a while back; well, we still do, but there are not many elm trees left. It is not actually Dutch, but came from America, in imported timber.

There were two trees I know that escaped it; one next door, although I had one a few hundred feet away that died even though I had it injected, and another a mile or so away. Both of these trees that survived had their branches inextricably mixed up with a pine and a horse chestnut tree.

A beetle that burrows under the bark, making a characteristic gallery, spreads the disease; they find the trees by smell. If you crush an elm leaf you will find that they have a very distinctive aromatic smell. I once had a garden table with the top made of a big elm plank, about six feet by three. It warped, as elm does, and after rain held about a quarter of an inch of water. One day I observed a couple of dozen elm beetles fly straight down into the water and walk about underneath on the wood. It seems they just can't resist it.

I wondered whether either the smell of horse chestnut, or that of pine, or both together put them off; they are both quite strong and distinctive smells. It would be interesting to do some experiments in an area where the disease is still spreading to see if it works. If so, perhaps it would be possible to isolate the compounds responsible and make a 'bug off' spray.

Dyer's Greenweed (*Genista tinctoria*) is a sort of small low growing broom plant. It is called Dyer's Greenweed because it was used to make dye, which is yellow, along with woad to make green, and was reputedly brought to England by the Romans. It makes seed pods similar to broom: that is, a pea-pod, rather elongated and black when ripe, but smaller than an ordinary one, standing out at about 45 degrees to the stem, pointing to the tip. It quite often favours coastal areas, and once, having landed by dinghy, my wife and I sat down near a patch of the stuff; it had black pods on it although it was early June. Then my wife said 'That pod moved!' On closer examination they were not pods at all, but little black bugs, a bit like caddis larvae with all their legs poking out from the case at one end. It was such good protective mimicry that it ought to be in the literature, but I knew it wasn't.

I then had to find out what the bug was, and nobody had heard of it. It looked as though it might be a bag-moth, or even a caddis that had hung itself out to dry before flying off, as dragonfly larvae do. This would have been most interesting. There are no salt-water insects other than mosquito and gnat larvae in brackish water; the crustaceans seem to have usurped their place. I don't know why insects don't get on in salt water; maybe there is something about chlorine or sodium that interferes with their metabolism. Maybe there is a simple insecticide waiting to be discovered. Eventually, I, or rather a retired professor of botany from the University of Wales, discovered it was a bag moth, last recorded in the 1880s: *Coleophora vibicella*. Strangely, a salt-water caddis was discovered a few years later, in Australia. It is enough to make one believe in synchronicity!

I wondered why it was out of synch with the plant, for Greenweed does not make black seed pods until a month or so later. I bet if you looked you would find the same bug in the Mediterranean region, where the Romans got their Greenweed, and probably the moth also by accident. Perhaps there they make seed pods earlier, the plant and the bug running two different clock mechanisms. It would be a most interesting study for a student to carry out; a bit of real amateur science.

I cut a tree down twice, once, if you can understand my English! We had an old overgrown orchard, and there was a sycamore seedling about eight yards high that was shading an apple tree. Sycamore seem to be spreading all over; I read once that they were introduced in about 1740, so I suppose the first few started spreading seeds around after the first two or three decades, and then a few decades later those started spreading seeds also. Now, a few generations on, they are swamping us. The last doubling of the series is as big as all that has gone before, plus one.

So I cut it down. It did not sprout the next year, but the one after that it sprouted up again and I left it, as it was not casting any shade on the apple. Seventeen years later I cut it down again. Now, I know it had only been growing for sixteen years, but when I counted the rings, there were nineteen! We had an oak tree that sometimes had an extra grow in autumn; the terminal buds would sprout and maybe make nine or ten inches, with rather small leaves. I wonder if that made a few extra rings as well?

It could be worth taking records from trees of accurately known age, to see if trees do sometimes shove an extra ring in; if so, we should perhaps statistically alter our tables slightly for climatic records and so on.

Mistletoe is a common parasite in Europe, principally on willow and poplar. It is less common in England but increasing, probably via imported mistletoe at Christmas; this could be checked as the continental variety is slightly different. I have seen it on lime, poplar, willow, apple and hawthorn, which is related to apple, in the UK. I have never seen it on oak, although it is said the Druids harvested it from oak trees. I have tried to get it to grow on my apple trees by making a nick in the bark and inserting a seed, but with no success; perhaps it has to pass through a bird first.

Once in the tree it spreads by sending a layer of wood under the cambium; this can spread a considerable distance. A red apple next door was heavily infested, with mistletoe twigs sprouting out all over. I have seen a lime fully garlanded near Winchester, but most clumps seem single. Some trees are impossible to graft together; it might be interesting to see if a piece of mistletoe wood could be used as an intermediary, grafting apple onto lime for instance, or hawthorn onto willow. There would be the risk of lots of mistletoe sprouts as well, but these could be easily rubbed out. There is perhaps some doubt as to whether it is fully parasitic or gives some benefit to the host; possibly, as it is evergreen, it donates some sugar in winter. At all events the host tree does not seem to be harmed, other than that a heavy load may make it more prone to blowing over. Comparative growth rates could be measured between two near identical young trees, one infected and the other not, to see if there is any difference.

Some ways of estimating age are a bit rule of thumb. I read that you could tell the age of a hedge by counting the number of extra species that had invaded the hawthorn or blackthorn it was made of. The figure given was one per hundred years. Apart from the fact that as new species get in, there are fewer new ones left, this seems to be a bit hit or miss; it might tell you more about bird species, or how near it is to a wood, than it does about hedges. I had a bit of linear wild land that I left alone; it grew five new species in a dozen years.

Another piece of revealed wisdom is a bit suspect. An estimate of air pollution can be gained by observing the amount of lichen growing on roofs, as pollution, particularly acid rain, kills it off. My slate roof is getting on for a hundred years old, but has very little growth. So high pollution then! However, I live in the New Forest, an area not particularly noted for air pollution. A kit hardwood garden table I assembled a few years ago has at least five different types of lichen growing on it, so maybe the type of slate or other roof covering is a factor to be taken into account.

Our previous house was near a cliff edge, and in autumn the cliff-martins used to wheel round the house, catching flies and stoking up for migration. Once one tried to take a short cut through the bay window, and broke its neck. As I bent to pick it up off the balcony, a large grey fly ran rapidly out from its feathers, towards my foot. I stamped on it as a reflex, but it was so leathery that it got up and staggered off. On closer inspection it had unusual wings; they were full length, but composed of only a strong rib, hooked at the end, and could not have been used to fly with. Later another similar fly made its appearance. I think I saw a glimpse of something similar on a BBC back garden nature program, on a swallow chick in a nest. It was not commented on, perhaps unnoticed, but a later program did cover parasites.

I found something similar once when pigeon shooting. One bird had flies, about the size of a bluebottle but grey and very flattened, under its feathers. These could fly moderately well, however. Perhaps birds that flock or re-use nesting sites are more prone to this type of infestation. It would be interesting to see if the two flies had a common ancestor; they were similar, but one further evolved than the other.

Some birds seem more intelligent than others, and are capable of learning. I have been told that crows in North America have learned to perch on street lamps, and droop their wings over the photocell to switch them on so as to warm their feet up. I occasionally used to throw crusts out, as one does; the magpies used to take them to the fountain bowl to soften them up. Even newts are quite bright. I used to throw bits into the pond for the goldfish, and the newts used to take a bit and hide it under a lily leaf before coming back for more.

My mother had a flat below us, and she used to leave one side of the kitchen window open in summer. The jays used to come in to fill their beaks up with butter, then fly out of the wrong window. Not so bright after all! An elderly grey-polled jackdaw got in once, and battered around knocking things over. It stared at me, perched on top of a wardrobe (the bird, that is), but allowed me to pick it up gently without a struggle, and carry it through to the next room while my wife opened the French doors. I gave it an upward swing and it opened its wings at the top of the trajectory, to glide off to a favourite perch. Although a wild bird, I expect it knew people better than I knew birds; it had probably observed humans for a decade or two, and thought them harmless, if a bit dim.

The local blackbird used to come in in summer, perching on the lounge window-ledge before walking twenty four feet down the lounge, turning right for six feet of passage, entering the kitchen to tidy up crumbs and then retracing its steps. It had probably seen food through the kitchen window, but knew it could not get in that way. The local blackbird here came into the kitchen once, but then went through into the bedroom, and tried to get out of the window. It flopped to the floor, obviously having broken its neck. I picked it up and it came to life; did it holler! A blackbird in an enclosed space is quite deafening. I laid him on the grass outside, one wing was spread out; one leg would not work and he lay on one side with his eyes half closed. His wife was most concerned, and came down to perch next to him. There was a lot of low chattering, totally different from the usual calls and song. Then she walked round behind him, grabbed a tail feather and leaned back on her heels. He flew up to the hedge. She knew her husband well, the old fraud! It still comes into the lounge when the door is open in summer, and perches on the chair arm next to me, not begging for food; it seems to want company.

I had a pet robin once, I did not really want a pet robin, but I fed it a bit of chewed up bread and butter once as it was obviously a lost fledgling, covered in fluff with a baby bird yellow beak. Every time it saw me thereafter it would perch hopefully for more, and it became routine. It lived for eight years, which is quite good for a robin! I used to put bits of bread on the bench, and it would always take the biggest first, to eat on top of the wall. Once, I thought I would give it a problem, and gave it two bits exactly the same size. It managed to pick up both at once.

If a Spanish airliner lands at Heathrow, it does not become British property, but still belongs to Spain, or, more exactly to the few hundred thousand shareholders in that particular airline. Even if I go up to it and shoot it with my air rifle, I cannot claim to have bagged it. However, somebody shot one of my birds in Spain the other day. How do I know it was mine? Well, I would recognise it anywhere; it had feathers, a beak and little feet.

In other words it was a kind of generalised bird, not mine specifically, but I had a part share with a few million other Europeans. One could argue that a bird is wild and free, and does not belong to anyone, singularly or collectively. Perhaps so, but a case could be made for the country of origin, especially if people have collectively spent millions on setting up reserves, long term studies and feeding in winter. Shooting does not confer the right of possession to the shooter.

If you are interested in wildlife preservation, you could get a couple of million signatures, I'll sign twice if you like; nobody is going to check, and anyway I might want to sign for each bird. You could then take a class action at the European Court of Human rights and sue for lots of money, at least a pound each. Perhaps it might be best to donate it to a wildlife preservation trust, or one or more in each country. You could tweet all your friends to get things moving. Some might argue that birds have no rights, and the court has no jurisdiction; maybe so, but humans do have rights and it inequitable that a few people should unilaterally deprive many others of the right to enjoy, and the many more commercially quantifiable benefits that birds bring.

I expect you will get objections on the lines that people will lose jobs in the gun industry, well, boo hoo! I don't really care that much. In any case, there will still be as much money circulating, and people will spend it on something else, so there will be no overall loss of jobs. You get the same kind of argument from the tobacco industry 'We have got to produce, or we will lose our jobs!' Tough; they know what they are doing. Some jobs could be classed as 'anti-jobs' inasmuch as they have a negative impact on society. You may come across the idea that you would be denying people the pleasure of shooting; well, let them shoot skeet, I say. A counter argument would be that the pleasure of a few should not have precedence over the pleasure of the many in observing and hearing wild birds, to say nothing of the job they do in removing insects, a source of great annoyance and damage to farming. Of course Spain is not the only offender; Malta is just as bad, and so are other Mediterranean countries.

Cats are strange creatures; a friend who stayed with us once had one, a rather plump tabby. After she moved to a cottage half a mile up the road, it used to visit; not us however; it was polite enough but did not wish to interact greatly, it just strolled into the lounge and curled up on the floor. A minute or so before its mistress arrived, it would get up and wait at the front door. It was unlikely to have heard the car coming, as when it first moved it would have half a mile away and behind a hill, also she did not always come by car. Stranger still, another cat she had, also a tabby, once met my daughter at the railway station, a mile and a half away, and pacing a few yards ahead, led her back home. They had never met before, my daughter having been away for a couple of years. Yet it knew what train to meet and that she belonged to the house. Finally, once after we had visited its owners cottage, a Siamese followed us when we left, yowling. It did not seem to want to come back with us, or for us to go back with it. I kept up a conversation until about halfway back, when something attracted its attention in the bushes, and cat nature took over. It died that night, of old age and kidney failure, so the vet said. I think it knew it would not see us again and was in some way trying to tell us.

So how could this be explained? One explanation is that a cat's mind is not so sharply focused as ours, we tend to have tunnel vision and to explain everything in words. Cats don't have words, but still have consciousness, wants and needs. We tend to think of the mind existing only in the head, for no good reason, other than we tend to think in terms of circuits, rather than field. It is possible that a cat's mind is more diffuse, being spread out in some degree over its territory. Native peoples think of their land in this way. They are not just figures placed in a landscape, but part of that landscape and the landscape is part of them. Similarly the cat's mind may be spread over a section of time; they are not clock animals. Thus a cat may be able to place its mistress geographically and chronologically in a landscape, and had already seen my daughter in the house tomorrow, and so knew where she lived. I had a nut tree once, and decided to pick it one night, the next morning. The squirrels got there first! They must have worked overtime. The same thing happened with my damson tree; one evening it was loaded, the next day we got only a pound or so. Perhaps they have a kind of misty view of a future where the tree is bare, and get in first. Alternatively perhaps the tree sends out a subliminal signal; 'Pick now!' I don't advocate this as a firm belief, rather as an explanation that fits the facts, while no other does.

Hidden toolboxes.

Within a plant's genetic make-up are hidden abilities, seldom used. For example, phragmites is a reed used for thatching. Most river estuaries have it and it was in the past cultivated in reed beds, although those are now mostly overgrown with willow. Reed for thatching often comes from Eastern Europe nowadays. Quite why it is cheaper to convey a bulky low cost item a thousand miles or more over land and sea I do not know; one suspects a hidden subsidy, or maybe we have forgotten how to cut it. Currently they are re-thatching the pub down the road with imported reed. There are twenty acres growing not a mile away.

It also grows on odd boggy patches, down cliffs. Every so often in such a situation it will throw a long stem over twenty feet long, which roots down to spread the patch. It does not seem to do this when the situation does not demand.

Bracken is another plant with similar abilities. I once cleared half an acre of woodland of bracken, as I wanted to establish a bluebell wood. If you want to do this, pull it up when the plant has made maximum expenditure in the new stalks, usually at about three or more feet high when the leaves are starting to unfurl. It will throw another set later, and again, but all the time from dwindling reserves. By year three you will have mostly cleared it. Be prepared to pull a few thousand stalks a day for about three months though!

If you want a bluebell wood, collect the ripe seed heads, just before they burst, from wild bluebells and scatter it, the stalks are dry and pull out quite easily. Scrunch them up in a bucket and broadcast over your wood. You will need quite a few bucketsful. The first year nothing much will show, unless you examine the ground closely, when you will find thin green tubular leaves a few inches high, rather like an onion. The next year a rosette of leaves more recognisably a bluebell plant will appear, and then next year they will start to bloom and seed off. Bluebells growing in grass often have a flattened rosette, hugging the ground, although some do not, so maybe there is a genetic difference. I don't know if these transplanted to woodland would grow tall; you could do the experiment. Bluebells are common in Britain, but not so in Europe, although I once found a small gorge in France that had plenty. There is a Spanish variety, which is more robust. Conversely, I once found a bank in France that was covered in a small pansy, which is rare in England.

With climate change, many species will have to migrate north or up mountains or perish. It might be a good idea for someone to set up an 'assisted migration' program, for many species might not be able to move fast enough.

Back to the bracken; one or two plants decided to move on, and invaded the orchard, travelling underground in dead straight lines, for thirty feet or so, throwing up a frond in diminuendo every couple of yards as they went. How they managed to keep the line whilst navigating stones and roots diagonally across a one in eight slope I do not know. Maybe they have some kind of compass.

Honeysuckle is another plant that throws extra-long shoots; these are more vigorous than normal, and if you section one you will find three times the number of vascular bundles.

Horsetail or equisetum is related to the coal measure plants. It grows in segments, with side branches at every whorl a few inches apart, each one smaller than the last as you go up the stem. If you pull them they unplug like plastic plants. Similarly the fronds unplug in the same way. If you section a stem you will find that it is hollow with five radial struts, like spokes; they are elastic and can survive in moving land, such as a slumping clay cliff. If I were designing a plant to colonise a new planet and reduce carbon dioxide, I would come up with something similar. One can almost see the underlying chemistry; I suspect a pentose sugar somewhere. Although very old they are not necessarily primitive; they have learnt a few tricks along the way. They have no parasites or diseases, whereas you might expect a very old plant to have had time to collect quite a few. Bracken is another old plant with no pests except a small moth that lives in southern Europe. It could be worth looking for a pesticide amongst their chemistry. Horsetail dies down every winter, so it is not frost resistant, except that if you constantly pull it to try to get rid of it, it will stay as a green tuft all winter. It too can navigate under concrete and tarmac, and is able to push its way up through tarmac by exerting a steady pressure. It also responds to threats; where a patch was partly covered by sand from a cliff slump, it threw out side branches that closely hugged the ground; pinning the mobile sand down.

It does not have many uses; it used to be used for polishing armour, so it might be good for the brassware, although both are a bit out of fashion now. Gold prospectors used to burn it to test for gold, for it cannot deselect heavy metals. It might be useful for cleaning up radioactive or other ground contamination. Best to test for concentration occasionally though; we don't want too much fissile isotope collecting.

I have an ordinary wild primrose plant, moved from a previous garden, which did nothing exceptional for a decade or two. The other year it went into excessive bloom; a foot wide hemisphere of flowers piled on top of other flowers, with not a leaf in sight. There was no room for more flowers, so it did the only sensible thing and threw up a nine inch stem with a cluster of flowers at the top, just like a primula, which is of course a closely related family member. Maybe it will morph into a primula and stay that way. I had another primrose that instead of making ordinary flowers, had the sepals expanded into green petals, with a miniature flower bud in the centre. We gave it to the Primula Society.

Are you involved in plant breeding or genetic transfer? Here is one plant that could make a billion. The mangrove is a tropical tree with high salt tolerance. It grows in dense monocultures along tropical estuaries, protecting and extending the shoreline. The dense root-mat protects small-fry molluscs, worms, crustaceans and anything else that gurgles and plops as the tide recedes, as large predators cannot penetrate it. Unfortunately large areas are being cleared for shrimp farms; so don't buy shrimps without knowing their provenance. The wood, although of small section, is extremely hard and durable, suitable for floors and doorsills. It used to be used for belaying pins, as it has a high breaking strain and resists rope burn. It also makes dense charcoal.

However it has no frost resistance. A temperate variety would have a huge sales potential as sea level rises. You only have to look at a map of the world to see the amount of coastline available. Admittedly some is wave-swept rock, but the coastline is fractal: that is, the closer you home in on it, the more indented it becomes. A brief glance at the Solent will show not a lot of coast on a large scale map, but you can multiply this nearly a hundred times of you explore the creeks, inlets estuaries and harbours. Take a look at Holland while you have the map out. You could sell tens of millions of trees there.

There is probably no one gene for frost resistance, rather a bag of bits. Most species have drifted in and out of frost resistance as the climate waxed and waned; there might already be some useful bit in the genome, inactive at present, or an active set in a closely related species.

The mangrove tree is capable of taking over. Once a thin green line has been established, it makes darts about a foot long, weighted at the end and with a couple of leaves to act as flights. These drop off, penetrate the mud and establish new plants, thus extending the line outward. That makes for easy planting too. Once established, it would be a good fuel-wood source, either chipped for a power station, or used for charcoal burning. However, don't do your burning in the old fashioned way, under a mound of earth, or even in a steel cylinder. Use something akin to a gas works, where it is heated in retorts, or a continual feed process via a worm drive. The gas can be used as an ordinary gas supply, while valuable products can be extracted from the tar. Some of the tar can be used at pressure to form briquettes from the chipped charcoal.

Musk plants, all over the world suddenly lost their scent in the nineteen twenties. No reason has been proposed. They must have been responding to some environmental trigger. Increased carbon dioxide is one that springs to mind. It would be interesting to grow musk plants in a low carbon dioxide atmosphere to see if they regain their scent. Of course, it may be that the plant has not changed, but our sensitivity to the scent has changed, so be prepared to live in your low carbon dioxide greenhouse for a while! At all events, there must be a trigger somewhere, and it would be quite important to see exactly what it is, for there might be other things to be triggered later as the atmosphere changes. What if wheat suddenly failed to set seed?

Octopi, relatives of the ordinary garden slug or snail, can be surprisingly intelligent. Offer one some food in a glass jar with a screw cap, and it will unscrew the lid. I once caught an octopus while skin diving off Malta. It was in an old piece of pipe like a four inch shell case. I brought it up to the surface, and it watched me closely from the security of its home until I broke surface, then quick as a flash it shot out, jetted down to the bottom and then ran over the rocks, its tentacles changing colour and pattern to match almost instantaneously. An octopus can solve a three dimensional maze by elongating its tentacles and exploring many avenues at once. I think I may be nearly as intelligent as an octopus; I was once presented with a transparent three dimensional maze, with a ball to roll through by tilting to solve it, at a Mensa meeting. I blew tobacco smoke through it.

Even the garden slug is quite bright; I have been told they can count up to three and read the *Sun* newspaper. They have a memory and can navigate, though; once, we had a ground floor room that always had slug trails across the carpet in the morning, which came up from below through cracks in the skirting, so I went in at night and caught some; they were about four inches long or more, and had dull orange longitudinal stripes.

I put them in a jam jar with some lettuce leaves until the morning. They did not eat lettuce, though, they lived on toast crumbs, but they wrapped themselves up in lettuce cocoons as it got light. I put them in the hedge opposite, but they had found their way back in a few days. The next time I put them half a mile away, down the hill. I expect they are hard at it still!

Bats can hear round corners! Once, I found a small bat hanging from a cornice on a mid-floor landing; it had probably been disturbed by some loft work the day before and got locked out. I thought to open a door and chase it out, so I walked through an archway, turned right and crossed a landing about eight feet square, down a flight of five steps to another landing, turned left, opened a door to a small lobby, and then opened an outside door. I did not need to chase the bat out; it flew out on its own account, having negotiated a complex geometry that needed four lines of sight. Presumably it knew its geography, and finding a lack of echo from around several corners, knew there was a way out. Echo location is a bit better than seeing, it seems. We used to hold barbeques, and as the dusk deepened the cliff martins were replaced by bats, one taking over seamlessly from the other. When I was young we used to catch them by tossing a small pebble in the air, whereupon they would follow it down to about three feet above the ground; some moths can hear bats, and close their wings and fall, but the bats know this. We would then throw a coat over and catch them; they bit our fingers. Nowadays you have to have a licence to handle bats. Should you think; 'Who would know?'; be aware that the last rabies death in England was a licensed bat handler, so, hands off!

So I present my scatter of odd observations. Not strictly science, but it could lead the way, coupled with other people's observations. Do join in if you have similar observations! You could set up a website as a meeting point for observers and experimenters.

Light relief:

Once, having bought boat tickets with my sister in law to Tuckton from Christchurch quay, we had ten minutes to wait for the boat. We sat down to wait on a bench five or six yards away, facing the quay. I was bemoaning the fact that I could not remember the name of a wall-growing plant (ivy leaved toadflax) that I had known for sixty years. My sister in law seemed very concerned (I think she also has the same problem.) She explained to me, being a librarian, that the memory is like a library; as content goes up, so does retrieval time. Mutually reassured, we got up to find that we had missed our boat.

Should you have a similar problem, I have found that deep breathing seems to help recall. Breathing oxygen might be better, or better still, breathing hyperbaric oxygen, for although blood saturation occurs quickly, tissue saturation takes a little longer, but don't try it for long, as it is poisonous at over one atmosphere pressure. So should you have access to a decompression chamber, you could give it a go.

SECTION 4

A QUESTION OF GRAVITY

This section is mostly about the universe. Cosmology is one of the most believed-in pieces of modern science, and also the fastest changing one. Here we have some new approaches to old problems. I expect they are mostly wrong, but I am in good company! There is a certain amount of redundancy, but as they are short pieces I have decided to let them stand rather than run them into one long article.

If I lift a brick from the ground and place it on a table, I have done work. The brick now has potential energy against the gravitational field of the Earth. If I push the brick off the table, this will be turned into kinetic energy, which will be converted to heat as it hits the ground. Energy and mass are equivalent: that is, the sum total of energy/mass is a constant. Therefore, there must be a slight increase in mass when the brick is on the table, because it represents stored work. This increase will of course be exactly balanced by the muscular work done in the conversion of sugar to get it there.

We can ask the question, 'where is this work stored?' Or, more pertinently, where is the increase in mass? Is it in the brick? In the earth? Or proportionately between the two? Or is it in the gravitational field?

The simplest explanation might be that it is in the brick, however, the work is really done by the field of the earth, the contribution of the brick being minuscule. Or we could say that it is stored by some distortion of the field between the two, but this gives us a gravitational field having mass, which may or may not be so.

Now, in considering a brick on a table, we are dealing with very small amounts of energy and an almost infinitesimal amount of mass equivalence. If the mass of the earth was concentrated at a point in the centre, like a mini black hole, we would find that at half radius, our brick would weigh four times as much, with a doubling of the kinetic energy available over the distance; each time we halve the radius, we get twice the energy. Thus our brick actually contains an infinite amount of energy, as does the smallest possible mass you can think of. This might be an indication that there is something wrong with the theory. It is possible that Newton knew this, but kept quiet about it!

If we propose a universe where one galaxy is receding from another at a good fraction of the speed of light, we have a very large amount of mass equivalence, and this could affect the attraction between them. It was only in the first half of the last century that it was realised that the universe was expanding, so any previous calculations regarding gravity did not take this into account, being based on a static field rather than a moving one.

This problem may be looked at in a different way. There appear to be two kinds of mass: static mass and an increase in mass occasioned by movement. For example, we can measure the rest mass of a proton, and then measure its apparent mass when travelling at a high velocity. From impact or deflection results we can show that the high velocity proton appears to have a greatly enhanced mass as compared to its rest mass. The question is, does this extra mass give rise to a gravitational field in the same way as rest mass does? Also, the energy depends on one's viewpoint; for a fly hitting a train, the extra energy is not great, whereas for a train hitting a fly it appears to be much more.

There is a further problem with mass moving in a gravitational field, which can be illustrated as follows.

If we approach a light source, it appears to be blue shifted. Although the speed of light is constant, we experience a greater frequency of peaks and troughs as we travel towards the source; also, the amount of energy we receive from the source is increased over unit time compared to what it would be if we were not moving. If we move away from the source, the opposite effect pertains and the light is red shifted.

Now, if a gravitational source is receding, its strength will decrease in relation to its increase in distance, but it is possible that it may also be attenuated by the fact of recession. In an extreme case we could postulate something receding at the speed of light, in which case we should not be able to detect any gravitational effect. The reverse situation would apply for a source approaching us.

In an expanding universe, where everything is receding from everything else, we should find that the effect of gravity is diminished according to the relative velocity of recession. We may or may not find that the increase in mass occasioned by this velocity exactly balances this diminution. Thus the force of gravity becomes conditional on two factors; the static mass, and its relative speed in relation to an observer. Doppler gravity. Why not?

ON GRAVITY

'Truth' is quite often a matter of consensus, and a new truth can arise if the arguments and evidence presented are cogent enough. Newton said all there was to be said about gravity, for a few hundred years at least, until Einstein, an amateur when he started his work, brought a new perception.

This paper, as much as anything, is to show a methodology, a way of looking at things, to raise questions that need to be answered. Being 'right' is not the object; it can be more important to be the first person to be wrong in a new field, as this can help others later. Quite often ideas birfucate, with the obvious result that at least one of them is wrong. It is however necessary to explore both avenues to form an opinion of the most profitable. We need therefore to speculate, the wider and more wildly the better, as by this process we will throw up ideas, some of which may be testable.

It has been said that everybody knows about gravity, but nobody knows very much about it. Quite possibly, at the very high temperatures in the beginning of the universe, it was at one with other forces, but 'crystallised out' as the temperature dropped, to be followed by electromagnetism which is in some ways similar, being a long range force. At all events, the third law of thermodynamics, which says that order increases at the temperature drops, seems to hold true in other instances. It used to be said that the universe was running down, or decaying to entropy. It is a funny sort of decay though; it seems rather to be running up to complexity, or decaying to information, itself a form of entropy.

Gravity of course is a force of attraction only, which obeys the inverse square law, and can be interpreted as a bending of space-time. The conceptual model of a rubber sheet like a trampoline, which is dented by an imposed mass, is commonly used; bigger masses make bigger dents. Smaller masses may be trapped in orbit around bigger ones, not having sufficient energy to climb out of the dent, but enough to circulate without falling to the centre. Light will also follow a curved path, although it has no mass and might otherwise not be expected to be deflected by gravity.

This model does not give the right kind of curve though. A better one might be a thin rubber sheet on the surface of water. Here the force of restitution is not only in the plane of the sheet, but also at right angles to it, so we get a meniscus-like curve, more nearly expressing the inverse square law. Also, with a very massive object, the dent can extend into a tube and then close up, giving a black hole model, where the connection to the surface is broken, something that will never happen with our trampoline model. Unless you are unlucky, that is.

Although this model is a useful alternative way of looking at gravity, it says nothing as to its cause. We might get an insight by looking at magnetic fields, as we know more about them. Superficially they are very similar; one can get a steel ball bearing to go in orbit around a magnetic pole; one can have a magnetic pendulum, have an upside down one if you like. They can repel though, which gravity cannot do. The field obeys the inverse square law. We could even look at the field as a sort of gravity that only affects iron, if all we have is a magnet and a ball bearing. However it is not the iron, but the electrons in it that give the effect. We don't need iron; a coil carrying a current will do just as well. Iron atoms tend to form octets, and the electrons in the outer shell tend to go round the lot, giving a circulating current. Many such domains aligned give the net magnetic effect.

If we have a superconducting magnet, we can bring a piece of iron up to it. The current will do work. If we then pull the iron away, we are now doing work on the current, and we can return the experiment to its original state. We can ask 'Where does this available energy reside?' We could say it is in the electromagnetic field. In a sense it is, but the circulation of electrons causes the field, so perhaps it would be more accurate to say the energy available resides in the motion of electrons, and this is expressed as a field.

The important thing to notice is that there are two states for the experiment; the initial one where all the energy resides in the current, and the later one where some energy has been extracted from the current. If we could measure our current we should find that doing work has diminished it; some of the current has been 'used up' or converted to another form of energy, which of course is exactly what happens in an electric motor.

Electrons are only a part of an atom; nearly all of the mass resides in the nucleus. Further, nearly all of the energy resides in the nucleus too. About a hundred thousand times as much energy can be extracted from the rearrangement of the nucleus of a uranium atom as could be obtained by rearranging its electrons by, say, oxidation. Even then, less than half a per cent of the mass will have been converted to energy.

All of chemistry is the study of atomic combinations caused by the re-arrangement of the outer shell of electrons. An electromagnetic field is also caused by the outer shell of some atoms, or by moving electrons not in combination with a nucleus. A word of caution though; the only reason why electrons cluster round the nucleus is the positive charge it carries, thus alteration of the state of electrons must in some degree be reflected in the state of the nucleus. The neutron, free of an atom and its associated electrons, has a half-life of about 12 minutes. It would be interesting to speculate on the effect of stripping many or all of the electrons from nuclei. Probably destabilisation would occur for some species, and a new route to atomic energy become available.

We can now take an inspired guess and say that gravity is probably something to do with the nucleus of the atom, to do with mass carrying particles such as protons and neutrons, or, at a deeper level, their constituent parts.

We might therefore expect a much stronger force, but in fact gravity is very weak for an individual atom, millions of times smaller than the electromagnetic force. Nuclear forces are very strong however, so we could suggest that they are very nearly in balance, but that a very tiny surface imbalance allows some to leak out. Or maybe we have a weak dimensional coupling, for if there are other dimensions they may be expressed to a slight degree in our more usual three, 'curled up small' so to speak. A very strong force in one or more of these dimensions would only appear as a weak one here. As an analogy, five miles of water in a deep ocean trench will only have the same surface tension as your cup of tea.

We get a similar sort of situation with photons; they don't seem to occupy time as we do. Ride a photon and you have instant elsewhere, by your watch. Ask a photon the time and it will tell you it is 'now' and now is forever. They shall not grow old, as we that are hadrons shall grow old. Just as well perhaps, otherwise we could not have quantum electrodynamics, our very best theory. Gravity also slows time, so we have another form of coupling. Gravity is cumulative however; we can't seem to 'use it up' in the same way as we can electromagnetic energy, it just goes on growing by addition with no antigravity to give a balance. The total gravitational energy of the universe is probably the greatest force around.

We will leave this line of enquiry now we have something to speculate about. It may be of course that magnetic and gravitational fields are totally dissimilar and one cannot use an analogy in this way. However, since they both probably came from the same parent source there must be common ground. If they recombine at very high temperatures, they must become the same, which is more difficult if they are chalk and cheese.

Next we will look at perturbations in a field. If we pass an A.C. current down a wire, we will get an oscillating field. Put a horseshoe magnet over the wire and it will buzz at mains frequency. Do a few more experiments and you might re-invent radio. If you use a D.C. current you will build a field, which will presumably propagate at the speed of light, but the inverse square law being what it is, not a lot of energy will be added to the field after the first picosecond or so. It may be that the growth of the field continually robs energy from the wire, or there may be some quantum barrier, where eventually there is not enough energy in the very disperse field after a few million miles, so that it just does not bother any more, not having enough energy to form even one quantum. We don't know, so we have something else to debate. Radio waves are perturbations in the field however, and they do carry energy. They are in fact long wavelength photons, although we normally think of photons as being visible light.

So what of gravity? If we could switch on a gravitational field, which we can't, we might reasonably expect it to propagate at the speed of light. If we could fluctuate the field we could have gravitational radio! We can't use photons however, so we will invent gravitons, and they will propagate as gravity waves that will stretch and squeeze space-time. The energy will be depressingly low however, given the weakness of gravity as compared to electromagnetic fields, so a big source like a black hole might be useful. Currently people are looking for just such waves, with no success as yet. If we could in some way hook them up with electromagnetic waves, perhaps find some sort of resonant frequency, we might have better luck, but I for one don't know how they might couple, or what frequency or overtones to look for. Given that they are out of the same hot stable however, there might be a chance.

We could however do some small-scale experiments. If we cannot switch a gravitational field on and off like a light, we can at least move one. If we take a large mass, say a ten-ton ball of lead, we can hang a smaller mass on a wire down a thousand-foot mineshaft, and bring it up close to the large mass. We should find a deflection of the wire as the small mass is attracted to the larger one. We could then replace our lead ball with two, joined by a stout girder, dumbbell fashion, and pivoted so they can rotate about a point midway along the girder

If we now leave our weight on the wire we can rotate the dumbbell so that first one, then the other ball comes in close proximity to our small weight. We ought to find that the weight is attracted slightly when in close proximity, but less so when the dumbbell has rotated ninety degrees, where the masses are further away and presenting a wider angle. It might be a bit slow, for the force would be very weak, but we should find that the small weight oscillates, and we could amplify this by finding a resonant frequency. Hey Presto; the first gravitational signalling! Perhaps a pointless and expensive experiment, but one never knows what might come out of it. Some bright person might suggest a way of using it as a carrier wave and imposing an electromagnetic ripple on it. We perhaps need to ask, 'if the two forces decouple as the temperature drops, how cleanly do they part? Over what range of temperature does this decoupling progress? Is there a halfway stage where we have a bit of one and a bit of the other? Do they come apart like broken glass, or more like Velcro? Are there any tricks we can play to feed a bit of one force into the other at normal temperatures? Can we use one field to influence the other, perhaps 'bend' it slightly?' If we can, we might be able to bend a field right round and turn it upside down! I expect we would expend as much energy into the electromagnetic field as we got out of the gravitational one, so not quite Cavorite, but better than rockets perhaps. Maybe we would need something exotic like two charged black holes rotating about each other to produce the effect, but maybe not.

We can take closer look at gravitational waves. They in some ways show a Doppler effect. We can use an analogy with light. For example, if we have two distant identical light sources, one approaching and the other receding, there are differences in the light we receive from them. The most obvious is that the approaching source will have its light blue shifted; the waves will be packed more closely together giving a higher frequency, while the receding source will be red shifted, if both are compared to a stationary source. Also, given the same rest brightness for the two sources, the approaching one will deliver more energy than the other, and it will increase in brightness over time. With an oscillating source we have something similar to a frequency modulated radio. An oscillating gravitational field will give the same effect; a bit more pull followed by a bit less pull, as in the rotating dumbbell experiment. A pair of neutron stars in orbit around each other might show the same effect.

However, what of the field? Supposing instead of oscillating, the gravity source is approaching. We could argue that an approaching gravity source pulls a bit more strongly than one that is receding. This could give a different value for the expansion of the universe; gravity would seem to decrease as one got further out, due to the high velocity of recession. After all, if a mass recedes at the speed of light, it is not there visually or gravitationally, so perhaps at nearly the speed of light it is nearly not there. We have another debating point!

Next we can look at how a field thins out over distance. There are two options; either the field thins out indefinitely, or it is quantised. The same argument was abroad thousands of years ago in relation to matter. Can you subdivide indefinitely, getting a smaller and smaller piece of a substance, or is there a limit, beyond which you no longer have the same substance, but its parts? Continuing, can you subdivide the parts indefinitely, or is there again a lower limit? Nobody talks of half a quark, although maybe they do have constituent parts; we don't know and are not likely to find out by experiment, as the energy required would be so high. There may be a mathematical route however, if anyone ever finds one. Long range forces can be thought of as stretching a bit of chewing gum; a lot of resistance at first, which rapidly diminishes as it thins. Short-range forces in the nucleus are rather like rubber bands; the resistance increases with distance and the band snaps before you can get out of the nucleus. Put in enough energy, and the added mass may give you two intact bands to play with! We can conceptually explore both routes however, and see if any differences appear.

First of all we can propose a universe which consists of just two hydrogen atoms, since they seem to be the building blocks of stars and so on. The field thins out to infinity, but we will stop just short of that so they don't take an infinite time to come together. They may therefore start off almost infinitely far apart, but will attract each other; eventually, all of the gravitational potential energy will be converted into kinetic energy and they will collide. They will probably just bounce apart again (neglecting valency forces, which we are not looking at here), but we can propose that they can radiate their kinetic energy load as heat and stay stuck together.

There are some things to notice:

1) The force of attraction is extremely small.

2) The velocity of impact will be small as this is a function of the kinetic energy available from the field.

 3) They will take a very long time to fall together.

So we have a finite amount of energy and a finite time. If we want to give a name to the unit of energy we can call it the hydrogen unit of gravity, or hug for short.

Next we do exactly the same experiment, but this time we have a quantised field. This field does not go on for ever and ever, but eventually says, 'Enough is enough; there is no more energy to form even half a quantum and I stop right here!'

Instead of an infinite field we have a field occupying a spherical volume of space. The field is finite as well as the energy it contains. We of course put our atoms right at the edge of the field, perhaps in passing wondering whether attraction begins where the two fields overlap, or where the two fields overlap their central atoms. The results of the two experiments are of course identical: the same finite energy, finite time, and finite speed.

We can go on though. Suppose we now have a universe with two pairs of hydrogen atoms stuck together, but each pair very far apart, and they start to fall together?

We re-run the experiment, twice. In the first case we have an infinitely extended field; you can put your atoms at infinity if you care to wait that long, but we can go just short of that to get a result. The result of course is twice the final energy, as the field was twice as strong.

We now try the experiment with a quantised field that occupies a finite space volume.

1) The field volume will be twice as big.

2) The acceleration will be twice the previous, acting over twice the distance, so we have a constant time of fall.

3) This constant time of fall would be the same for any identical pairs of masses falling together, and would in fact be the age of the universe at greatest expansion, if it is 'flat'; i.e. if it expands and stalls, neither going on expanding or re-contracting.

4) The final velocity will be twice that of the previous experiment, as we have twice the acceleration acting over the same time.

5) The kinetic energy will be four times as much, as we have twice the final speed.

So we have two different results from the two different assumptions. Which one is right? We have something else to debate.

We can continue doubling up. We can have two lots of four hydrogen atoms, and let them fall together. Then eight, sixteen, thirty two, and so on. We will very soon get into very high numbers.

Some more problems arise here. Firstly, we have a rise in kinetic energy whichever model we choose. Eventually the closing speed will be so great that the mass of the atoms will begin to rise. This is interesting. We have unlocked this huge potential gravitational energy of the universe, and it is being converted into kinetic energy, which starts to pile up as additional mass. It seems that we have two kinds of mass; we can call them static mass and kinetic mass. We must ask the question 'if all this potential energy is stored in the field or the atoms that create the field, does it have mass? Is an atom at a far distance from a gravity source the same mass as when it is resting on the surface?'

If the bits of the universe were flung out and dispersed in some great explosion they must have had kinetic mass, which has disappeared as they slowed against the gravity field. So where has the mass or energy gone? Under conservation laws it must be expressed somewhere. Or perhaps the kinetic mass is the same as the difference. Perhaps static mass does not exist, but is really the expression of some internal movement within the parts of the nucleus? All mass might be a result of energy of motion.

If we split a uranium atom we say that about a half of one per cent of the mass is converted to energy; the sum of the parts is less than the whole by what is called the packing fraction. However the total energy must be the same, so perhaps if we added the static mass of the parts to the kinetic mass gained by them moving apart at high velocity, we would get the same mass as the intact atom. In another analogy we could have a very strong flywheel rotating at great speed; it is not going anywhere, but possesses great internal motion. We can ask whether a rotating flywheel weigh more than a static one. So we have some other things to debate.

Another problem is that if we go on increasing the field indefinitely, we reach a point at which light has its wavelength decreased so much that it cannot escape; we have a black hole, so we are told. However, even for a black hole the amount of energy available from falling into it is finite, if very large, yet the energy to escape from it is supposedly infinite. Something does not quite add up here. Similarly, if we have a space ship just on the event horizon we are told that it would have to have an infinite mass of fuel to escape. Actually, it has; one advantage over a rocket is that the fuel is moving with you and carries kinetic energy. Take the case of two identical steam trains, racing and accelerating down two equal tracks. One is a ton lighter and is slowly gaining on the other. The gaining driver realises that he is short of a ton of water however, and drops a scoop into a water trough, picking up a ton. They are now the same weight, but the energy cost of picking up the water and accelerating it to train speed has slowed the gaining train down. Moving mass is worth more than static mass.

There is a problem with a rocket climbing out of a black hole however. By the time it gets out, the universe will be over, for time stops at the event horizon, while the universe evolves at an infinite pace in comparison.

There is a lot more mileage in gravity. We could compare a gravitational field with an accelerational one and find quite a lack of equivalence, despite Einstein, or again with a centrifugal one to find more differences. We could do a lot of work on spin, the most fundamental and least studied property of the universe; we don't really know if it is a thing in itself, or if it can only exist in relation to something else. We could separate a mass from its centre of gravity, that point at which the force of gravity appears to act, with a gyroscope. We could explore hollow planets with no gravitational field inside the hollow; only that mass below you attracts. For the mass above, all forces balance. We could extrapolate this to conceptual 'shells' within the universe, again only that bit below you attracts, for the universe is not attracted towards its centre. All the separate bits attract each other and the net result gives the appearance of gravity acting towards the centre, which is not the same thing. Or we could drill straight holes through the Earth and drop things through, finding a constant time of transit (47 minutes) wherever we drill our hole to, be it just down the road or straight through the centre and out the other side. We could play billiards on a peculiar table, divided in the middle by a time line, where time flows at half speed in one half compared to the other, and find out how mass and kinetic energy appear to change. We could compare the concepts of zero field and zero net field, for a stationary rope lying across a meadow is in a different state from the same unmoving rope with a tug of war team on each end of it. For gravity, the time flow might be different, so there are lots of things we can do. Better still, there are lots of things you can do.

Enough for now however. The important thing is to ask questions, and then to get others to kick ideas about; the more rigorous treatment can come later, it is fun and it is proto-science, for the questions must come before the answers, also anyone can have a go.

ON THE DISTORTION OF
GRAVITATIONAL FIELDS

A spherical gravitational field surrounds the earth, diminishing by the inverse square law as one moves outwards.

Where another body such as the moon is present, the two fields reinforce or diminish each other according to the position of a reference point in the two fields. For example, between the earth and the moon, on a line through their centres, there is a null point where there is no net field. This will be at a distance which is the ratio of the two masses, about one to one sixth in this case. This null point is actually a point on a null plane; were the two masses identical, this would be flat and extend indefinitely equidistant between them. As the earth has the stronger field, the null plane is bent, having a parabolic section curving away from and beyond the moon. Also, the field on the other side of the earth will be enhanced by the addition of the moon's component, proportionate to distance.

However, there is another element to be introduced, for the moon and earth rotate about a common centre of gravity. This gives rise to a displacement between the actual centres of gravity of the two components and the points where they are apparently acting. This is really a form of refraction, although the term is more commonly used for light.

To illustrate, consider a sailing boat, sailing at right angles to the wind. The apparent wind experienced on the boat is not however at right angles, but displaced forward, being compounded of the boat's motion and the movement of the wind, giving rise to a simple parallelogram of forces. For example, if the boat and wind speed were the same, the apparent wind would be displaced to 45 degrees off the bow at a speed of the square of the sum of the square root of the boat and wind speeds.

Thus, for an object moving in the gravitational field of another body, which can be considered stationary, the apparent gravitational field will be shifted forward, and appear slight stronger than the static field. As a gravitational field is assumed to propagate at the speed of light, this effect will be slight. It may be possible to detect it however, if the apparatus is sensitive enough.

Such an apparatus might be formed by a very long suspension pendulum in a vacuum, being held above the rim of a fast rotating massive flywheel. If for example the suspension is 100 metres long, the movement of the bob by a few millimetres will require an extremely small amount of energy, being that required to raise the bob by that diversion from a straight line occasioned by tracing an arc of a circle a few millimetres long on a radius of 100 metres. Response might also be rather slow.

If the bob responds to the gravitational field of the spinning flywheel being at an angle to that of an unspun wheel, we have an example of gravitational refraction.

In setting up, it might be wise to use two identical pendulums, for a pendulum as described will not stay still. The passage of the moon and sun overhead will affect it, for the earth suffers a tidal effect on land as well as on water, the crust tilting, rising and falling as the moon goes over. There may even be some gravitational lensing effect at a solar eclipse; this might not be simple, but something akin to 'Newton's rings' may be produced. The Allias pendulum experiments seemed to show more than a simple addition in the way that the fields of the moon and earth reinforced each other at the time of eclipse. There will also be other sporadic events, such as earthquakes and even traffic; these will not be of great significance unless there is a periodicity that coincides with that of the pendulum. A number twenty-seven bus might do it if the timetable is regular! Thus the movements of the reference pendulum could be subtracted from the test pendulum, to give a net effect. If the pendulum bobs are reflecting spheres, a very weak laser could be used to measure displacement, as a very long optical lever could be arranged. The laser should only be used for a short time, as this will give a slight push to the pendulum, or two opposed beams could be used.

Another experiment might be to use two flywheels, one on top of the other, one moving, the other not. We might find that the spinning one, in the fullness of time, drags the other round with it, due to the gravitational angle presented. Alternatively, we might find that the stationary wheel contra rotates, due to the enhanced gravity presented by the approaching mass as opposed to the receding mass. We should beware of electrostatic effects however, and also moving electrons relative to each other will give rise to an interaction even if there are no free electrons present. It might be necessary to compare flywheels of different materials and different electron configurations to eliminate this effect.

Our experiments will be low speed, but events in space can move a bit faster. An object travelling at an appreciable fraction of the speed of light may exhibit Doppler gravity, that portion of the field ahead of it suffering compression. Thus the near passage of such an object might give rise to a gravitational boom, the field rapidly rising, then falling as it came to nearest approach, this effect being in addition to that rise and fall expected under a static field model as the object approaches and recedes. To give another example, the flight of an arrow can be approximated by a series of still snapshots. However, an infinitude of such snapshots will still give zero kinetic energy and not truly represent the whole picture. In the same way, a moving gravitational field is not the same as a series of representations of a static field.

These considerations will affect the calculated expansion of the universe, the expectation being different from that calculated from a static field model. A departing object may attract more weakly due to the diminution of the field occasioned by its movement, while an approaching object may attract more strongly due to field compression. In the extreme case, for example, an object travelling at the speed of light will have no forward field. Such speeds are not possible, but very high speeds of a considerable fraction of the speed of light are.

QUANTUM GRAVITY

If offered the choice of interest payments yearly or monthly, it is likely that you would choose the latter, the reason being that you would receive slightly more, the difference being the interest on the interest over the period. London Interbank Borrowing is at a daily rate, while the best of all worlds would be interest at such a short interval that it becomes virtually continuous.

Classical gravity assumes a continuous acceleration by way of interest; however, if gravity is quantised, the continuity might break down for small masses in extremely weak fields, such as a gas cloud condensing to a galaxy.

Over time, the two models might diverge noticeably, for in the quantum model acceleration would proceed as a series of discrete slight tugs, with a period of inaction in between, rather as a photographic plate or a squid (a superconducting quantum interference device, not a relative of the octopus) may collect discrete photons to slowly build up an image of a distant star. We would be back to yearly interest as opposed to continuous payment. The principle observable differences would be that acceleration by gravity might be lesser at long distances due to this 'delayed payment' effect. Also, there would be a mismatch between the general collapse of a cloud and stellar nuclei formation within it, the latter proceeding more rapidly in comparison due to the stronger or more continuous field pertaining. Very low-mass objects would tend to fall more slowly than heavier ones, in contravention to Gallileo. This sorting effect would lead to friction that would enhance halo formation, the reduced kinetic energy being productive of smaller orbits and less escape for dust grains forming the halo.

Is there anyone out there who could calculate the value for a graviton, using any known effects of this kind, or conversely, using a range of values, derive possible outcomes that could yield testable predictions?

A GRAVITY WELL

I dusted off one of my Gedanken centrifuges the other day, to see what alternative use it could be put to; one has to be green, and recycle and all that.

First I ran a few simple experiments. I put a tube from the centre to the rim, suitable for sliding in a very fine chain I had; it will flow like water, but unlike water, you can pull it out again. After spinning up, I started to slide the chain in, down the tube. As it went further down the tube the pull on my end got stronger, so I wrapped it round a reversible little electric winch, wound as a dynamotor, and found that I could get the centrifuge to do work. I noted that while the chain was moving, it was pressed hard on one side of the tube as it was accelerated to the speed of the centrifuge at that radius, but not at all if not moving, only the tension from the artificial 'g' being apparent. Eventually I had put quite a lot of chain in and the centrifuge had slowed considerably. On measuring the loss of speed I found it greater than expected from the amount of energy extracted, but this was explained by the increase in mass occasioned by the addition of the chain. There were two elements to consider, the relative position of the mass within the centrifuge, and the energy required getting it there. If you have ever wondered why an ice skater speeds up so much when changing from an arms outstretched spin to an arms held in spin, with a relatively small ratio between the masses concerned, that is why; there is work added also. I then pulled the chain out again, using energy via the winch. The net result was that the end state was the same as the first state. Gedanken experiments don't have frictional and hysteresis losses unless you dial them in. You don't have to double up with an identical experiment at 180 degrees round the rim to balance the thing either.

I simplified my centrifuge by ignoring the spin, merely representing it as a gravity well; you can think of it as a convenient tin can you can stand on a bench, with the unusual property that things weigh a lot more when put into it. Given the fact that the gravity increases as to the depth of the can, there is anisotropy if one moves about the field; putting things in weakens the field, and there is an initial energy cost to set the thing going. The approximation works well enough.

I put a tube down my gravity well, across the bottom and up the other side, switched on, and noted that it sucked a lot of air in as the field built. Temperature sensors showed that the air was very hot at the bottom, due to the compression. I puffed a little air down the tube, finding very little resistance, which was surprising considering that the air went down one side, increased its pressure and temperature several hundred times, then reversed the process up the other side of the tube to appear as normal temperature and pressure air at the outlet. This was a completely reversible result. I then extracted some of this heat, using first a silver bar, to conduct it out, and then a tube filled with water at high pressure. The bar was a bit slow, but the thermal siphon took heat out very rapidly. I blew a little more air in and was surprised to find it difficult, the air in the exit tube being a lot denser now, due to the extraction of heat bringing the molecules into a closer pack and therefore weighing more. I used the heat extracted to run a small steam engine. Unfortunately my Gedanken centrifuge is only a qualitative device; it can't do numbers, so I don't have proof that the heat energy extracted exactly equals the extra energy need to push the gas through. The gas coming out was a lot cooler too; the thermal gradient was much greater than for a normal refrigerator, and any old gas could be used. For an industrial process requiring a hot and a cold end, my centrifuge would be just the thing!

I then decided to use my gravity well as a chemical reaction vessel. With a suitable catalyst I could make nitrogen compounds from ordinary air, the only energy cost being that of pumping the denser cooler gas up the tube and out. I then tried electrolysis of water. The device siphoned well as the bubble rose, once the water had become saturated with dissolved gas, and I could extract energy from the flow; however, I discovered that it takes more electrical energy to electrolyse water under pressure, but could not quantify. Hot water electrolysed more easily, so if you have to electrolyse water as part of a process, use hot water; some waste heat can reduce the electrical energy required. All kinds of chemical processes requiring pressure and heat could be used; for endothermic reactions a pumping load is required, and for exothermic ones energy can be extracted by way of the favourable pressure differential. I then put a small diesel engine at the bottom of the tube, driving a generator to extract energy, and found I could get a very high compression ratio and consequent improvement of efficiency, plus a bonus from the reduced pressure in the exhaust due to its elevated temperature. I tried running an air motor, pumping air at pressure down the down side, but allowing expansion at the bottom to give energy via a dynamo and the expansion reducing the load on the up side. Energy was generated, but I could not quantify this to see if it equalled the pumping load. It should do, but only if the input and exit temperatures are the same.

What is needed next is a computer model that can quantify, and compare costs with the normal way of compressing and heating gases. I think a simple centrifuge might be a lot cheaper than multistage piston, compressors and burners to achieve temperature rise: only one moving part!

CLOSE ENCOUNTERS

Neutron stars are Nova remnants, which need a critical mass to form. Part of this mass also needs to attain a critical density, at which electrons are forced into protons to form neutrons. The energy to overcome the Pauli exclusion between electrons is provided by gravity and the recoil pressure occasioned by the acceleration of the outer layers of the star as it explodes.

Free neutrons have a half-life of about 12 minutes, decaying to a proton, an electron, a neutrino, and some radiant energy. If two neutron stars underwent a close encounter, tidal effects could disrupt them so that material was torn off both stars. We have no direct experience of the characteristics of such material, but it would probably be a fluid of low viscosity, which would form droplets which might go into orbit, fall back into one or other of the stars, or escape the gravitational field of both. Such a mass would be sub-critical: relieved of the massive pressure inside the star, neutrons would be free to decay as normal. In addition, depending on how close to criticality the stars were initially and how much mass was removed, a situation could arise where one or both of the stars went sub-critical.

Such an event should give a clearly identifiable signal, a characteristic of which would be an explosive flare up period, which would depend largely on the initial conditions, such as mass, velocity and separation of the stars. This initial flare might be short, as closest approach speeds could be in the region of hundreds or thousands of miles per second; there would then follow a decay period with a characteristic twelve-minute half-life.

Evidence could be looked for in old globular clusters, old because there would be time for neutron stars to evolve and globular clusters because the ejection of fast light stars, by slingshot effects, plus the slow reduction in kinetic energy due to tidal effects between stars, would lead to closer packing. Binary stars are fairly common, which could be evidence for close encounters. It would be interesting to examine old records, to see if such a signal has been observed.

It is interesting to consider the case of two black holes, in orbit around each other. Orbits tend to decay due to tidal friction; water movement in the case of the earth, and rock deformation for the moon. However, for two black holes there is likely to be no loss, for the deformation of the two event horizons, there is no friction.

There are several ways the two black holes may interact. First we can consider two black holes in the same plane, perhaps the most likely arrangement. If rotating in opposite directions, the two edges will rotate together, like two cogs in mesh. Quantum pairs are ejected infrequently from a black hole, but there is a much greater density of pairs, which never escape, near the surface. The distortion of the field may lead to a greater escape of these. As the orbital rotation is likely to be very fast, we should get a flashing beacon, like a lighthouse, if we happen to be in the same plane. If rotating in the same direction, we will get a different signal.

Another mode of orbit is where the two event horizons, which are likely to be disc shaped, are stacked one on top of another initially, and rotate at right angles to their spin. Thus they move to a similar position as in example one, then the bottom one comes on top, and so on.

A third mode of orbit is where one is rotating as in the first case, but the other is vertical to it, rotation around the first in that plane. Again a different signal should be produced.

N.B. For the sake of brevity there are some omissions of detail in the above. I will expand the detail in the form of brief notes, as follows:

1) The twelve-minute half-life should be adjusted for recession velocity; the signal would be stretched, as the latter part would be delayed by having a greater distance to travel.

2) Electron capture by protons is pressure dependent. However there is a pressure gradient in a neutron star, and the critical pressure could be below the surface, with some heavy elements floating on it. This would affect the outcome of a close encounter. We could perhaps envisage a dynamic shell at the critical depth in which some neutrons are decaying, while others are being formed. For ejecta,

there would be a possible route for heavy element formation.

3) Neutron stars are likely to have high rates of spin. The rate, radius and alignment of the two spins would also affect the outcome of a close encounter. Equatorial speeds might approach escape velocity if radio signal blips show the true rate of spin. If neutron material has a very low viscosity however, the star might continue to ring like a bell for some considerable time after formation. It is difficult to model a cause for the ticking of a neutron star. Electromagnetic radiation presupposes electrons, or, as a last resort, protons; neutrons seem an unlikely source. Also, it is difficult to envisage some radiative anomaly on the surface of a ball of neutronium; it ought to be undifferentiated. In addition, if we find a rate of ticking where the spin speed is in excess of escape velocity, then the ticking is not due to spin, and we must look for another cause. Residual volumetric oscillation might be a cause, but we would expect the amplitude to decay, not the frequency, as found. There must be a loss of kinetic energy, and this suggests material being ejected at over the equatorial spin speed, or a tidal couple with satellites.

4) Black holes also have a critical mass. There is a possibility that two black holes, close to their critical limit, could be similarly gravitationally disrupted into evaporation, for the only way to distort a gravitational field is with another one. Effectively, there is a point between two masses where the gravitational fields cancel, with a reduced gradient as one approaches the null, and a neutral plane around the null in which a mass is attracted to neither centre of mass, but tangentially to the neutral point. This plane would be curved if the two masses were not identical. There is the potential for 'orbits' in this plane, but they are all unstable, decaying to elliptical orbits around one mass, or a figure eight around both. These orbits would be further perturbed by the mutual orbits of the two stars about a common centre of gravity. If the concept of a zero point acting as an attractor seems contra-intuitive, remember that the net field at the centre of the Earth is zero, as it is, for that matter, at the centre of a black hole.

Whether this net zero field is identical to no field at all is a moot point; it would be interesting to fly a clock in a Legrange point to find out. A signal from a black hole disruption would be sharper, and have a higher energy than that for a neutron star.

1) An energy audit for neutron star formation might be useful. A nova represents a period of large fast changes in energy distribution. Gravitational collapse provides kinetic energy, some of which is absorbed by creation of nuclei heavier than iron, some as electromagnetic radiation and some to provide escape velocity for a considerable fraction of the star. The rest, which remains in the neutron star, is either used up in forcing electrons into protons, or in forcing the spin. It is perhaps not commonly realised that when a spin radius is decreased, not only is the original kinetic energy of spin conserved, but also an additional increment is necessary to overcome the 'centrifugal' pseudo gravitational field. To give an example: if an ice skater spins, holding a couple of ten kilogram dumbbells at arms-length, and then pulls his arms inwards, work is done to achieve this, and this work is stored as kinetic energy in the spin, in addition to the initial spin energy. Thus neutron stars are likely to be ellipsoidal rather than spherical. If they undergo further collapse to a black hole, more gravitational energy will be transferred to spin, so the probability is that a black hole cannot collapse to a point, but to a circle, like a string or a loop of gravity. In fact all black holes must have spin, for all stars do. Although gravitational forces tend towards infinity, so also to centrifugal forces; we have in effect two opposed fields. This would lead to a zero net field null in the centre, in the form of a circular plane rather than a point. There might be the possibility of some interesting orbits through this null, for the event horizon would be torroidal.

2) Two tidal waves would occur for each star, one leading, where matter is pulled off from the body of the star, and the other lagging, where the star is pulled away from the wave. The two waves would not be mirror images of each other, as the leading wave would be nearer to the centre of mass of the other star and therefore larger, the ratio of the waves being related to the ratio of the radii of the stars and the distance between them. This is also true of the Earth/moon couple, but here the ratio of Earth radius and distance from the sun is much larger, leading to a very small difference.

ORBITS IN THE NEUTRAL PLANE

Quite often maths has no application; this is the best sort of maths, curiosity driven, free of all commercial drive. Blue skies and all that. The trouble is that nobody much wants to pay actual money for it. It is no good telling them that it may come in handy later; they are of the 'where there's muck there's brass' school.

So therefore we should ignore them; let them employ number crunchers to crunch their numbers, I say. We can't actually say our maths is useless, instead we make it elegant. We could say it had inutile beauty, al la Nabukov, in the margins of our undeserving journey. A turn of phrase, by the way, derived more as a result of clumsy translation from Russian than skill in English. Now, beauty has its own reward; ask any pretty girl. It may be possible to do some maths that is elegant, and, perish the thought, can be sold as well.

If we propose two equal gravitational masses, there is a point midway between them where a third mass will not be attracted to either. We can say there is no field, or that there are two opposed equal and opposite fields. The difference may be more than semantic, for gravitational fields affect the rate of time flow; we will not pursue this here however. We can call this the neutral point; it would be very easy to represent this on a computer display, if a trifle dull.

We could also place a mass, still equidistant from the centre of the two masses, at a short distance to one side of this neutral point, whereupon it would accelerate through the neutral point, then decelerate past it to reach an equal and opposite position on the other side. It would then cycle indefinitely. Actually, not quite indefinitely, for the system is unstable: the slightest deviation from the exact neutral would cause it to fall to one or other of the two masses, it being in effect a variant of the three body problem. Any point equidistant from the two masses would give the same effect. The totality of these points we could call the neutral plane, meaning that on this plane a third body would experience an equal attraction to both masses, and therefore not fall to either, following instead a resultant path. For two equal masses the plane would be a flat surface extending to infinity.

Now this oscillation about a path equidistant from the two masses is in fact a species of orbit. To demonstrate this more clearly, we can add a little energy at right angles to the path, a sort of nudge sideways, whereupon the path will open out to an ellipse. Add more energy, and we can open the ellipse out to a circle, yet more and we have another ellipse where the major axis is at right angles to the first. It may seem odd to think of a straight line as a species of orbit. For a single mass, one would have to drill a hole straight through the centre. Perhaps it might be better to say that all orbits are elliptical, with the extremes being a straight line and a circle. It was, after all, in the sixties that a computer invented a new triangle, which had two equal sides of half the length of the third, one angle of 180 degrees, and two others of zero. This of course could be represented graphically and dynamically on a computer screen, which is a little more interesting.

Next, we can get a little closer to the real world by using two unequal masses, like the earth/moon couple. Here, the neutral plane is deformed into a parabolically curved surface, and we can perform orbits on this in the same way as before. This is a bit more visually interesting; we could show the surface as if it were a sheet of graph paper, perhaps distorting the size and shape of the squares so they represented an equal transit time, or perhaps energy value for transit.

Next we could add rotation; after all, if we have two gravitational masses they would fall together in the real world, unless they were in orbit around each other. The orbits would now become much more complex: some might self-destruct while others would become stable. We might find a few new LaGrange points, the surface becoming a volume, plus some that looped round both bodies in a figure eight fashion. We would in fact have a full-bodied three-body system, so to speak. We could run a series, adding energy slowly to show points of instability where an orbit suddenly flipped from one stable configuration to another. We could even add other optional masses. I once met a mathematician at an open day at UBC, in Vancouver, who had a three-body system up and running. He was trying to get the two larger bodies to eject the third. He asked me if I thought he could then go on to eject one of the two bodies left. I had to tell him that it was extremely unlikely; they can lack common sense, these computer buffs!

I suggested he model a globular cluster, a sort of small nebula without spin, with a few million stars. (Well, you could start with a dozen or so.) Here we have a multi body problem; they don't collide much, the whole being mostly empty space, but they do form binary stars, capture being a result of energy dissipation due to tidal effects at close encounter, which are quite common, and triplets. It would be interesting to see if there are more triplets in a cluster than in a rotating nebula, like ours. I think there ought to be, as in a spiral nebula capture is restricted to a plane. The point about a globular cluster is that stars are ejected; it is like evaporation, they carry away more than their fair share of energy. The interesting question is, does the rest of the cluster contract, due to the reduced average energy, or expand, due to the lesser gravitational mass? It may depend on initial conditions. I expect he is still beavering away at it! Personally I think that evaporation must lead to condensation.

All in all we could devise a very pretty screen saver, perhaps where the user could drive it by altering variables. We would have to patent it; of course, and see if we could sell it. It would be pirated of course, so we might not get as much money as we would like, but never mind.

HALF A PHOTON

Let us suppose that we have an extremely fast shutter. We could make one out of two discs, with an aligned pinhole in each, contra-rotating at very high speed. The two pinholes would align for a very small fraction of a second, to allow a single photon through from a photon gun, firing in synch with the rotation. The pinholes would align again half a rotation on, which could be useful for reflection, double slit experiments, or for a reference beam. The proximal disc could be silvered to reflect out of synch photons back. We could then slowly retard the photon gun, to the point when cut off occurred before the photon had time enough to get through.

Half a photon! Except that there is no such thing, photons being quantised; they have a fixed amount of energy, and a fixed wavelength. It is true that they come in a range of energies, so we may have a sort of second level of quantisation as well if space is quantised, maybe in Planck units. The range would then probably be finite, for there is an upper limit in a finite universe: the energy cannot exceed the total energy available, while the wavelength cannot be shorter than that determined by the total energy, which may coincidentally be the Planck length. It is unlikely that a wavelength longer than the diameter of the universe could exist, unless it is twisted, although we might have a virtual extension of the wave out to half a wavelength beyond the boundaries of the universe. We can show this by setting up a right angled prism, with a ray of light entering at right angles to the long face, being internally reflected off one short face across to the other, and exiting in parallel with the incident beam. If we bring another prism up to the face giving the first internal reflection, at about half a wavelength the light will start to be transmitted instead of reflected; once in contact it will all be transmitted, surface inequalities apart, and there will be no double reflected beam. We can do the same thing with radio waves of a few metres, and presumably the universe could 'sense' another one within half a diameter. So maybe a shell of virtual possibility having another seven times its volume surrounds the Universe.

So what might happen? Perhaps there would be some kind of tunnelling effect, as with tunnelling electrons. Maybe half would be transmitted and the other half reflected, or a varying ratio according to how much of the photon got through, or maybe we would get two photons of half the energy, which means they would be of twice the wavelength. That leads to the question of how the length extends. Does the front end go faster than light? Or does the tail stay connected to the pinhole until the correct length is attained? Either way we still have half a photon for the time it takes to sort itself out.

We do a get a similar sort of indeterminacy in the double slit experiment; here the same photon can appear to travel though both slits at once, so presumably half goes through each. The problem may be with the way we tend to picture a photon. The photon does not inhabit time and distance as we do. For the photon time has stopped, and the universe is flat. Thus there is no before and after for the photon, nor any distance in front of the slit or behind it. For the photon the problem does not arise; it does know its ends and its beginnings. From its point of view it travels nowhere in zero time. The beginning and end of the universe are all one to its own perception of events

However, we have a different point of view. If we could generate two photons of twice the wavelength we could see what happens if we recombine them. Do they have a memory of what they once were? Probably yes, as in the double slit experiment, in which case we may have a much more robust entanglement than mere spin. We could also cross polarise the two halves; double entanglement. Spin could be put into the mix as well. Very good for quantum computing, for instead of yes, no, or maybe, we have two more factors to play permutations on. Quantum computing cubed.

We could reflect the photon or its fragments back and forth through the two pinholes, shredding it into smaller and smaller energy packets. Photon dust, or more likely wool, for we would soon have very long wavelengths. The modern equivalent of philosophers' wool perhaps?

ON THE EXPANSION OF THE

UNIVERSE

Distant galaxies show a red shift, which increases with distance. This is interpreted as an increasing recessionary velocity implying an expanding universe. Since this expansion appears equal in all directions for the visible universe, we could assume we are at the centre. This is statistically improbable, so we can assume that the universe is bigger than the observed portion, and that, rather like dots on an expanding balloon, the expansion will appear to be centred on the observer wherever he is. An alternative explanation is that light is in some way robbed of energy in transit, but there is no plausible mechanism proposed, or a sink for the lost energy. Light does of course spread out over distance, leading to a reduction of energy per unit area, but this seems to lead to a reduction of amplitude, not wavelength.

There are two possible modes of expansion, both of which could act at the same time, or serially. In the first mode, the universe origins in the form of an explosion, with a high initial velocity which should be diminished by gravity as it progresses, either resulting in a subsequent implosion, or an ever diminishing rate of expansion if above a critical value (where the energy of expansion exceeds that taken back by gravity), or a trend to stasis, which will never be reached.

This transfer of kinetic energy to potential energy against gravity will affect the density distribution of matter in the universe. If we compare the initial explosion to that of a chemical explosion, we have an original state, a solid in the case of an explosive, which contains energy locked up by some mechanism, which is released very rapidly. Expansion proceeds from the outside, because it is only free to do so there initially, giving rise to a temperature and density gradient. This will be modified by deceleration due to gravity as the outermost atoms are slowed, leading to a trend to an increase in density as one proceeds radially. These two trends are in opposition and the exact gradient will be a compromise between the two, according to initial conditions. No such gradient is observed. Maybe there is no gradient, or maybe the observed volume is too small for it to be apparent. If so, we could derive a measure giving the minimum size of the universe necessary for it not to be observed, but no maximum. In both the expansion of a chemical explosive and the universe we need an initial volume. A point cannot expand without the formation of more dimensions; two will give a shell, with a hollow centre and three the observed universe. We can of course propose more dimensions.

The explosion model will not explain the increasing rate of expansion observed, so we need an additional explanation in expanding space. In this model space itself expands, carrying matter along with it, rather like the dots on an expanding balloon. An early hyper expansion has been proposed; there are two possible ways in which this may affect the observed expansion rate. Firstly, the expansion of space may 'kick-start' the universe, and, when over, leave the galaxies with a residual motion caused by the expansion. There is a problem here in that we have to propose a coupling mechanism for the expansion, where space does not just expand around matter, leaving it in situ, but carries it along with it. Later we have to propose a decoupling mechanism, when the expansion or hyperexpansion has stopped, so that matter does not lose its velocity of expansion, rather like the dots on a balloon when one ceases to blow it up.

The expansion, in either mode, implies the creation of new energy, or its transformation from another source, by virtue of the increased gravitation potential available from the greater separation of matter.

There may be a problem with the way we interpret velocity with red shift. Currently we assume that the light from a receding source is red-shifted, which is well and good. However, the light is climbing out of a gravity well, so even if the distant source were not moving it would also experience a red shift. This may or may not be compensated for by it entering out gravity well, but the two fields may not be equal, and in any case we are not at the centre of ours.

Again, if we examine a clock next to our source, we will find that it is running slow, due to the gravity field it is at the bottom of. Thus we have another red shift not caused by velocity. Also of course, a receding field may not have the same strength as a static field at the same distance, as noted previously.

Time dilation by a field is due to red-shifting. For example, if we set a clock giving a time base signal on a neutron star, the pulses will be spread out, causing it to run slow to a distant observer. Put the clock in orbit, and we have 'oscillating time'; time running backwards for part of the cycle! You won't believe this, so I will leave you to work out why.

Personally, I have a problem getting to the other side of zero, but I could be wrong!

SECTION 5

A TRIP ROUND AN ULTRAFUGE

Here we explore some rather more outlandish devices that are not so likely to be developed just yet. I use the fiction of seeing them in operation in a parallel universe, as this makes them a bit more believable.

One advantage of visiting parallel worlds is that one gets to see technology that does not exist in this one, which gives the opportunity for technology transfer. The big advantage of information is that is transferable, while the handful of gold coins one may seize in a dream are not even dried leaves on awakening. Information has no mass and no energy and is in a sense implicate, inasmuch as it can be instantaneously transferred anywhere, provided you have the means, that is. But don't think the streets of an alternative universe are paved with gold; they are paved with chewing gum, the same as anywhere else. You get the same thing in religion: I was once told that it is virtually impossible to get something out of the Kingdom and into this world; presumably it has been tried.

An ultrafuge is a super centrifuge; we have very good ones here, capable of pulling half a million 'g' I am told. They are used for precipitating small molecules in solution, and separating isotopes in a gas. The one I had a look round was several orders of magnitude more effective than this.

We entered a building that seemed to have rather high security, which I won't detail: I noticed a neat little analogue computer on the way in. It used the intersection of curves, the circle, spiral and involute (like a section through a snail shell), for two dimensions, showing a continuous gradation from a circle to a straight line then a helix, and three other curves I did not know the names of, being a helix that increased its pitch, another that increased its radius, then a combination of the two. I'm not sure how it actually worked, but digital computers rapidly run into awfully long strings of noughts and ones; analogue can be much faster, and, who knows? Maybe there are some things that cannot be reduced to noughts and ones! It has been said that we have traded our wisdom for knowledge, our knowledge for information; maybe we are now trading information for texting. It used to be information overload, now it is garbage overload.

At the heart of the lab was a hemispherical construction about twenty five feet across; we entered through an airlock. There were various cautionary notices. 'Do not enter when running light on'. 'Interlocks off'. 'tixe on'; with some of the letters backwards, and half a dozen lights which were all green. There were enough optical cables to run a telephone exchange for a small city. We suited up,

"For the dust," my guide explained. "Although we do have electrostatic precipitation as well, we have a one way system with the airlocks, one in, and one out, like an hotel kitchen, as there are sensitive pieces of apparatus to be carried in and out, so we don't want a collision!"

The centrifuge itself was in the form of two 'coolie hat' circular devices about a yard across, one inverted, they would be closed together when running, but were open now. My guide explained:

"The shape is because it is laid up from multiple strands of very high tension material, a sort of semi diamond structure carbon nanotube. And, of course, as they come to the centre, it has to be thicker. There are some circumferential strands as well; more to do with the construction, to keep it in shape, for although you could lay up circumferentially, they have more mass for the same effect than the radial ones. The small cavities near the rim are for experiments. It is very important to balance the apparatus when running, so we do experiments in pairs. An ordinary weight will not do, as the test material may alter its centre of gravity under high load. The bearings are magnetic and the whole runs in a hard vacuum. One problem is the time to pump down, actually pumping is fast enough, but then we have to do the gettering."

"What is gettering?" I asked.

"Removal of the last traces of gas, called gettering, not because it gets, but that was the name of the inventor of the process. We use a probe at near zero to condense the last molecules, and pre-flood with hydrogen, as it is a faster molecule that comes out quicker. Also, for the odd ones left in, they are of low mass when they do hit the centrifuge. Activated gas carbon used to be used, as it captures molecules, but it is too slow."

"Isn't the kinetic energy of different gas molecules the same at a given temperature?" I asked.

He looked at me as though I was a bit slow.

"It is not the molecular speed that counts - that is only a few centimetres a second at the temperature we run at," he said. "But the centrifuge speed."

I noticed that the whole centrifuge was gimballed, presumably so as to avoid coreolis forces caused by the rotation of the earth. There was also a very thick ring of steel, not rotating, also gimballed.

"That is in case of a detonation," grinned my guide. "At the speed we run, the energy of the rotating mass is many times that contained in the same weight of dynamite. Also, the floor would blow down and the white hot gas go through a quench, which is rather low tech, being a room full of gallon water bottles stacked on Dexion racks. The initial steam explosion is rapidly cooled as it goes through more and more water, the whole exiting down an eight foot wide octagonal tunnel we got from a redundant coal mine explosion lab, which points into a grass bank."

I had noticed this peculiar long structure on a plan of the site at the entry; there were also two more, spaced evenly down the complex.

"The ring is not just armour plate though; we have magnetic containment as well. Around the outside of the centrifuge is a cryogenic carbon fibre electromagnet, S pole out, and another around the blast ring; S pole in. The field wraps round like a Tokomak. The result is a magnetic squeeze, which means we can exceed the tensile strength of the centrifuge material."

There would be another effect, which he did not mention; if you put a soft iron rod in a coil, and switch on, the rod will contract, so there would be a magnetic tension as well. If you don't believe it, put some iron filings in a raspberry jelly, and try that.

He handed me a small slug of metal, about as thick as the sort of pencil you get in the back of a diary. "That is the kind of separation we can achieve."

It seemed odd in the hand, one end much heavier than the other. The light end was a silver colour, fading to purple in the middle, then gold.

"What is it?" I asked.

"It started out as a gold magnesium alloy, 50%. But, as you can see, the 'g' field has pulled them apart."

"How can you stop the heat leaking out of the molten metal and warming up the centrifuge?" I asked, for the melting point of gold is over a thousand degrees C, and even with alloyed with magnesium, it would still be over 800 C.

"It was cryogenic," he replied.

That rather took me aback. Exactly what 'g' do you need to pull an alloy apart? I felt a bit out of my league; this was serious stuff! Maybe it was a wind up however. The purple bit in the middle seemed a bit odd. I had not heard of a purple gold alloy; you would expect a paler gold, similar to nine carat, being somewhere between silver and gold. After all, you would think it would be used in jewellery.

Also, why use magnesium? There are several lighter metals. Potassium, sodium and calcium might present a problem if passed hand to hand, but beryllium might do. I don't know much about its alloys however, and it is a bit brittle. Metals are a bit picky as to what the mix with. Tin, for example, alloys with copper to make bronze, and also with lead as in pewter. Zinc also alloys with copper to make brass, but if you try to mix it with lead it merely floats on top. It will also pull out any silver in your lead at a ratio of three hundred to one. So if you are melting down any old church roofs it might be an idea to check, particularly in the Peak District. There was a firm there during the war offering new roofs for old. Some of the old ones were up to three per cent silver, as the ore was mixed and they did not know how to separate them then. They may still have problems; a guide to some of the workings told me that not long ago they had to send some ore to Canada for smelting. "Too much zinc," he said. I wonder.

Actually, if you are into church roofs the thing to look for is 'low alpha' lead. Freshly mined lead always has some radioactivity, principally due to radium caused by the decay of uranium, which is present in small amounts. Refining removes the uranium, but not the radioactive by-products, which are mostly various isotopes of lead anyway. Over the centuries the residual radioactivity decays, giving 'low alpha' lead. This is in demand for shielding sensitive experiments on cosmic rays, for example, where stray radiation can mask any sought after events. Archaeologists complain bitterly that old Roman anchors, coffins and pipes have a habit of disappearing, due to the high price of low alpha, which is hundreds of times more than ordinary lead.

"We do other experiments," he went on. "I will show you the results of 'frame drag' where the rotating mass drags space around with it to a small extent."

"Do you get magnetic field drag as well?" I asked. He looked at me sharply.

"Yes, we do," he said, but did not expand.

I had noticed a sensitive pendulum to one side of the 'fuge. It consisted of a very fine suspension cord, probably silica, with a small gold ball on the end. Next to it was a box, about a meter long, labelled; 'Multi reflection optical lever, 1 Km. Temperature balanced, 0 to 40 degrees centigrade.' There was also an optical thrust correction table, for various wavelengths.

"It is far too sensitive for here," my guide explained. "It was designed for something else; even the wind blowing on the building affects it, and people walking about; so we leave it on all the time when running and take an average."

I had a plumb bob with a needle point, and thought that was rather good, but they seemed a bit ahead of me here; an optical lever a kilometre long would find atoms a bit large.

"We also do some experiments with light beams in opposed directions."

I knew a bit about that from a different source. It looked to me as though they were also looking at both sides of a collapsed probability state, but I did not let on, after my query about magnetic drag. I was only a guest after all, and if they did not want to tell me it was not for me to complain. Exactly what result would you get if you move a South Pole past another South Pole; very quickly? Not a lot, I would have thought, but there would be a shear, which might lead to eddy formation, particularly if you spaced gold or tungsten weights round the rim to increase the local gravity field. You might be able to achieve resonance and amplify the effect, which would be exceeding small. A kind of cavity gravitron valve perhaps? We are quite happy with the idea of a high 'g' field inside the centrifuge, but what if we could bring a bit of it out? After all, if gravity is merely bent space, why not bend space the other way to achieve lift off? Very useful for getting things into orbit, but not so good between the stars; no gravity there, or at least only the weak pull of distant stars, which is equidirectional anyway.

Of course, since gravity slows time, reversed gravity might speed it up. I will leave you to explore the ramifications. It would explain why flying saucers can appear and disappear. At all events, they were spinning fields like a candy floss machine.

I realised they could, probably did, have electrostatic compression too. On top of this there would be very high 'g' forces. Exactly what one would get from this I do not know? Light is a combination of electric and magnetic fields, screwed up together; add gravity and there might be another particle, light with mass, maybe? Anyway, you do not need very high security at the door for pulling alloys apart. We do not know enough about gravity. We assume that the gravitational field of the sun penetrates through the moon at eclipse, with no interaction. However if there is no interaction, how does it 'know' the moon is there? Bent space solves that problem, but raises another, inasmuch as that any other fields are bent as well along with the space; so gravity does affect magnetic and electrostatic fields, if only by bending the space they are in.

There are questions to be asked about field interaction. For example, suppose we have a coil, which can be energised by a current to form an electromagnet, and a few feet away a soft iron bar. If we switch on the coil, the bar will also become magnetised. This is due to the iron atoms in the bar being in the form of octets, with some electrons circling the whole, as though it were one big atom. In the soft iron they are arranged randomly, producing no net field but under the influence of the field, they arrange themselves head to tail, giving a net magnetic effect. Presumably the field of the electromagnet expands at the speed of light. Supposing we could slow this down and observe the spread. At first the field would propagate in all direction from the coil; eventually it would reach the soft iron and start to attract it, giving a force that could be measured. However, the soft iron would also become a magnet; and its field would propagate in all directions, eventually reaching the coil; thus there would be another increment of attraction. So we have a two-step process.

If we substitute a bar magnet for the soft iron, we have an existing field which would presumably start to interact with the field form the electromagnet as soon as it was switched on, the force reaching a maximum once the electromagnetic field had reached the bar magnet. This is a different situation from the first experiment, although the 'steady state' end result might be the same if the magnets are of equal strength. There are permutations with other fields which I will leave you to examine. Presumably a gravitational field behaves in the same way, for although we cannot switch on a gravitational field, we can move one. Thus the moon has an 'observed position', which is where we see it, and where the apparent gravity field is centred, and also a 'real' position, about a mile ahead, to allow for the time take for the field to propagate. This lag might lead to a twist in the field, which would tend to bring the moon slightly closer, and speed it up very slightly, giving a slight difference in the apparent force or gravity. This would not be much in the case of the moon, but might be considerable for a star with a neutron star companion. Magnetic fields can also be wound up in the same way; the sun's equator rotates faster than the poles, leading to the magnetic field being twisted; eventually this 'short circuits', leading to sunspots, and magnetic storms on earth.

There were other experiments to find out if light is bent in the proximity of a rotating mass, it will be, as a result of the mass, very, very slightly, as in gravitational lensing, but they were looking to see if there was any enhancement.

I won't continue, but this was an excellent example of a multifunctional device, where many different experiments can be performed on the one machine. The moral of this tale is, when preparing your pitch for funding for a very expensive bit of kit, it would pay you to look at what else can be done with it. Funders like a bit more bang for their buck! The big ring at Geneva is a prime example.

As a cautionary note, when proposing 'Many Universes' for problem solving, you will find that many people don't agree; they are of course in the wrong universe, but you can't tell them that. Amongst the physics community, particularly quantum computer buffs, you more or less have to believe. Also philosophers (starting with Leibniz in 1709) have the same kind of covert acceptance, so you may get more change there. Everett was not the first: I knew about it when I was seven, quite a few years before him. And Dr Pangloss thought he was in the best of all worlds, although many of his clones did not agree. Don't worry about the restriction of free will that is inherent; of course you are able to change your world, but for another that is equally deterministic! If it all gets too much for you, you can always skip to a world in which alternative universes don't exist!

In one sense they have to however; just as time dilutes things so they are spread out (ask a photon for the alternative). Multiple universes do the same thing; another sort of time perhaps.

It also explains the Big Bang: after all, if time is condensed so that everything happens at once, and space condensed so that everything is in one place, and all the alternative universes heaped up together, it is no wonder that a bit of expansion took place! I wonder it did not happen sooner.

Mathematicians don't have a problem; also they will tell you that not all possible worlds have to exist. It is quite possible to have an infinite number string in which the number one does not occur. Rather unlikely, you may think? But you can have an infinite number of number strings; you will find one amongst that lot! Several, probably.

For the time being it is best to consider it just a dream in the head however; after all, your image of this universe lives there, so why not a few more?

ON MOVING ROCKS

In the fossil record, one telltale sign of early man is worked flints. Some of the Old Stone Age artefacts such as hammer stones may take a bit of recognising, for a stone that has been used for hammering can look rather like one that has not, apart from a few significant flakes off at one end. However, Old Stone Age points such as arrow and spear heads are instantly recognisable, as also are knives and scrapers after a few have been found. A manufactory may contain tons of reject splinters, while a 'flint mine' such as Grimes Graves shows extensive working over a long period of time.

This is not to say that some level of culture may not have existed before the Stone Age. There may well have been social structure, the loom of language, the use of fire, an interest in the moon and stars at night, and wood working, if only by breaking and burning points. Weaving and plaiting, to make rope or string with the possibility of a kind of loop or sling, might well have predated flint or obsidian in some areas, but these have not survived, or their survival has been embraced in later culture and thus re-dated. New Stone Age worked flints are much more sophisticated; many more, smaller flakes have been taken off to give a more graceful tool, while many are polished as well as knapped. There is more art embraced with the functionality; perhaps a prettier point meant a higher status.

Nor is man the only tool-making animal. Chimps may strip a leaf stem to poke into a termites' nest to tempt them out to be eaten. A tuft of leaves may similarly be used as a dip to get water from a knothole far up in a tree, while they can be readily taught the use of human tools such as bicycles, even if they could not make them themselves. Crows will also strip leaves to make a repeatable recognizable tool to extract grubs from under the bark, and several birds use thorns for the same purpose. Sea Otters will use an 'anvil stone' on their chests to crack molluscs, while elephants will sometimes surprise their mahouts by using a bit of wood as a lever when handling a difficult log.

Social insects make complex structures, which if not tools in the normally accepted sense, represent 'fixed plant' by way of a housing complex which is highly organised for diverse functions. All of insect behaviour and much of mammalian and avian seem to be instinctive. Even beaver dam building seems to be so. The European beaver was hunted nearly to extinction and ceased to build dams as numbers declined. As they regained numbers they also regained dam building, yet there was no unbroken cultural link for this.

Perhaps there is some kind of instinctive drive which needs to be interpreted with intelligence. In humans there is probably a greater use and range of intelligence, but one should not reject the instinctive element. Much of human behaviour, while showing great ingenuity, is not really intelligent if considered in relation to the apparent problem to be solved. A look round the shops at the vast range of inutile manufacture will prove the point. Much of it is an attempt to deceive the purchaser into parting with their money rather than serve any useful purpose.

Later, large stones were moved. Early stonemasonry shows the use of very large blocks weighing many tens or hundreds of tons. Early South American stonework is noted for large random blocks, as found on the ground, worked so that they fit together exactly. Early Egyptian work is the same, as is also early Greek. One little feature that you may find amongst this use of random stone is a straight joint, perfectly flat and even, at an angle of about sixty degrees to the horizontal. There is one in the 'mortuary temple' near the Sphinx, another in the gatehouse wall at an early Greek city in Albania, and I have seen them in photos from South America. I can predict that there will be one somewhere in Troy and any other major sites from around that millennium.

I know nothing of their significance. Perhaps they are some kind of mason's mark, or a religious symbol. Perhaps the angle varies with latitude; at all events it is a deliberate use of a straight line and angle where there are no others and there must be a reason. Also, their use raises the question, 'if they could cut blocks accurately, why did they use the more fluid form of making each block fit the next according to its shape and dimensions?'

The answer is probably that it was far less work, using the tools and skills they had, to make the best use of a block without the labour of dressing it plane, and in the process throwing a fair amount of it away. The obvious conclusion to draw is that as stone working skills improved, it became more economic to use a standard size and weight of stone that could be easily handled. An alternative explanation may be that as the ability to handle blocks weighing hundreds of tons was lost, early Man was forced to spend more labour in cutting smaller blocks. Some of the weights were massive. Many people have marvelled at Stonehenge, where many blocks weighing tens of tons are accurately arranged after having been transported from Wales. However, at the temple to Baal at Baalbeck (later Romanized) in Iran, one of the blocks weighs a thousand tons. It is not the ground course either; it is laid on top of other blocks weighing five hundred tons, although I have read a different source which leaves the thousand ton block in the quarry, and elevates the others to seven hundred and fifty tons.

The Egyptians used blocks weighing thousands of tons to erect stelae, like Cleopatra's Needle on the Thames embankment, only much bigger. We can see exactly how they cut them out, for there is one half finished in a quarry in Egypt. The method is quite simple. Mark out, peck away with a hammer stone or possibly a bronze tool until you have cut a trench to the required depth, then undercut until nearly free, after which one rather carefully props and chocks, taking care of fingers and toes until the block can be finally broken out; after which you can do the fine finish. After that, all that is left is to put it on a reed boat capable of taking a thousand tons, and float it at flood time to the nearest river to its destination. Easy, they did lots! Strangely, the one half-done is flawed; there is a crack running through it, which is probably why it was never finished. It may be an earthquake crack, occurring when it was half done, but it does not look quite like that. My theory is that this was a training post for raw recruits, before they were let loose on a good one!

There are obvious questions as to how early people handled such large masses. By the time the pyramids were built a standard block of about two and a half tons was used for much of the construction, with bigger ones on the thicker courses. Some of the blocks inside the pyramid weigh nearly a hundred tons; they however were special cases, where a big block was needed. Any football team could handle a two and a half ton block with a bit of organisation and a few primitive aids.

One can only assume that bigger blocks were moved by bigger teams. Rollers were possibly used in some cases, and sleds lubricated with some liquid are pictured, with lots of men dragging. If one used a one in twenty slope, the direct pull to overcome a thousand ton weight would only be fifty tons, plus another fifty perhaps to overcome static friction. Probably two or three men per ton would do the trick. Exactly how one handles a stele, a stone needle a couple of hundred feet long, and then gets it vertical, is still a mystery. To start with, taking the length/cross section into account, it would be as brittle as glass; jack up one end and it would crack across the middle. Also there is a problem with tipping it upright. The bottom corner would spall as the weight came onto it, for no way can one put a weight of a thousand tons onto a stone corner only a few feet long; splinters would fly off like shrapnel. Possibly they were produced with a rounded 'rolling edge', which was cut off later. That is the way I would have done it, if I ever took the job on in the first place. Probably I would use a stiff 'hydraulic mortar' to bed it down into a shallow socket, and have lots of ropes and sightlines to make sure it was vertical. Exactly how to get it from the horizontal to the vertical without breaking it or losing control I will leave for others to work out. A wooden frame would not help much as wood is far more flexible than stone. Having a thousand years or so tradition and experience, starting with little ones, probably helps.

Sometimes people wonder how the Aztecs fitted stones together so accurately, without the use of mortar. Actually, it is no great problem. All you need is a tool made from three stout sticks, as long as your stone is wide. With these you plait in a large number of other sticks of equal length, as if you were starting a hurdle, so that you have something like a two sided rake. If you place this on top of the stone and push the protruding sticks down until they meet the surface, you have a mould of the surface of the stone; on the top surface of your rake you of course have a reverse pattern. If you then place it over the stone to be fitted, you can chip away at the surface to be worked until it corresponds with the line of the sticks. In practice you would probably work both surfaces to minimise the amount of stone to be dressed off. To fit the sides of two stones together, as well as the horizontal faces, would require a bit more planning, with the correct offset used so that it all fitted. After a few expensive mistakes, and some re-working, one would soon get the hang of it though. The advantage of such a fit is that it is pretty earthquake proof, much more so than cut masonry blocks. The lintels of Stonehenge are located on pegs sticking up from the uprights, fitting into holes in the lintels. The tallest one has an obvious peg, but I don't know if the capstone is still around. Maybe they should put it back; the Royal Engineers put one capstone across two uprights a while back. So the builders knew about earthquakes too. Probably the use of lintels in Egyptian temple architecture is for the same reason. I have seen it said that they used this form of construction because they had not yet invented the arch. This is not true; they used arches in the smaller tombs of lesser mortals. The problem with the arch is that earthquakes hit you sideways; if you hit an arch from the side the top strikes up, and on the return the deformed arch is more likely to crumble than a stone lintel, which may joggle and grunt a bit, but stay up.

Of course the arch is much more efficient in the use of mass, and can have a much longer span, as everything is in compression. If I were using lintels I would make them wedge shaped, wider at the bottom than the top, maybe with a belly sagging down as well, so as to put much more stone to spread the tension load, but nobody seems to have thought of that. Similarly, if I were designing Gothic cathedrals, I would do away with all those flying buttresses, and make the two side walls lean inwards, and curve them slightly from end to end, so the two walls were similar to two flattened clamshells leaning into each other. Nobody seems to have done that either. The Knights Templar seemed to have been involved in the cathedral building craze that swept Europe. They had got some plans they thought were of Solomon's temple, although I would have thought that would have had lots of lintels. At all events, we seem to have had an even newer stone age for a while. If you look at Kings College Cambridge, built by Henry VIII at no expense spared (he had just robbed the church of a hundred times as much), you will see that they did quite impossible things with stone. If I had not been told otherwise, I would have thought it was designed by a chef/sculptor skilled in working with icing sugar.

We sometimes forget that this was the Stone Age. Everybody talked about stone, and went to see that latest rock. Tribe would compete against tribe to have a bigger stone circle. If you wanted to be important, you just had to have some rocks; there was absolutely no other game in town. Even the provinces joined in. Maes How in the Orkneys has the finest bit of corbelling outside of the great pyramid, while Newgrange in Ireland, with its white quartz facing dominated the surrounding land, combines grace with mass.

We call the next age the Bronze Age, but the first use of bronze was probably for cutting stone. Egypt was a Bronze Age civilization, although they actually used iron too. Iron rusts rapidly in the soil of Egypt, which is rather nitrous; small artifacts would have rusted if there were any, so we may have a low count. There are some rather massive iron clamps used in one of the early pyramids, though, and in the Great Pyramid the ventilation shafts have little iron doors, gold plated, near the ends. A robot has been sent up, and photographed one. There seems to have been surprise at this; however, in the last century someone tipped some explosive down the top of one such shaft and blew out a little iron door. They were told it could not have come from there as Egypt was a Bronze Age civilization!

I would not use expensive robots. I would use bats. The pyramid is full of them: three species! All one has to do is to catch a few outside and give them a small tag, like a transponder bar code, that could be read at a distance. Then one could see if one could find them again when they had roosted inside. If most of them have gone missing, they know something you don't. Similarly one could identify where they emerged and see if one could find them inside in daytime. If there are any secret chambers it is likely the bats know of them. One interesting feature of the known galleries and chambers is that they are offset to one side of the centre line. The builders must have had a reason, and there are a variety of explanations, probably all wrong. One possible one is that there are other chambers which would have interpenetrated if they were both on the same centreline. You can see the same problem if you sit on a flat twin motorcycle; one cylinder is ahead of the other, because they have to clear a twin throw crankshaft. The obvious solution is to displace both sets of galleries sideways so that they miss each other.

Another way of looking for hidden chambers might be by pumping air. When the pyramid at Dendera was opened, it was very hot inside, until the workmen broke through a wall and a stream of cool air flowed for several days; an event never repeated. The obvious thing to do would be to block up the wall again and pump air back down the other way; then, when released, one should get the original flow back again, and this time trace where it was coming from with little smoke tell-tales. If the original report is accurate, then a considerable volume must be connected. This might be a naturally occurring cavity. The reason for the pressure differential might be seasonal, or hydraulic, where air was trapped at a slightly higher pressure by an 'invert', a water trap that would prevent air flow up to a certain pressure, according to the water height; you have got one under your sink. Other very large spaces have been found elsewhere, and there is reference to an underground city which may be a myth, or may not. Cosmic rays have been used to X ray part of the pyramid, but this will only work for a cone above the detector.

Other things to be looked at while out there, are the rates of erosion round monuments. There is no record of the building of the Sphinx; actually it was not built, but a few hundred thousand tons of rock excavated around it. The blocks you see are a result of repair work, three or more thousand years ago, which is recorded. Someone has noted that the pit it sits in shows evidence of water erosion, as the rain poured over the edge. This means that it must have been dug twelve thousand or more years ago, for it has been dry since then. Also the head seems to have been re-cut at some time; if it was originally a lion head, then it would have faced the rising of the Lion at about that time. Similarly, the erosion rates for the pavement around the pyramid don't add up. The pavement has eroded down below the casing blocks, but the erosion rate does not match the erosion from when the casing was stripped off in places. However, there would have been considerable run-off from the faces while intact, whereas once stripped, water would have tended to soak into the exposed core blocks, giving less run off. One could check the erosion at the centre of a face against that at the corners, for run off would have been less there.

Pyramidology is great fun. One can speculate endlessly. Newton spent time in Egypt, as he wanted an accurate measurement of the circumference of the Earth for his theory of gravitation. He had read that the baseline was a quarter millionth of it. He was thwarted by the fact that the base was obscured by thousands of tons of rubble and sand though. He should have known that the area of a face was a couple of acres however, and been able to work out the baseline from that, for if one knows the angle at the top of an equilateral triangle, and the area, then one can work out the length of the sides. Perhaps he did. Mercator was there too, and left his graffiti. He was interested in mapmaking. Not surprisingly, the Mercator projection is exactly the same as the one used by the Egyptians; i.e. the world as it would appear from one earth radius (4000 miles) above the surface. If you like measuring things you will find that the cubit is half a yard, and the Egyptian inch is the same as the Roman one, but the British one seems to have lost about a thousandth over the years. If you are engaged in arcane subjects, such as looking for the lost Ark of the Covenant, then you will find that the dimensions as given in the Bible are exactly the same as those of the coffer in the King's Chamber in the Pyramid. So perhaps as well as God, the Ark contained a system of standard measures as well; quite useful for a nomadic population. It would save a lot of squabbles in the market place about exactly how many yards of cloth and how much wheat was agreed for a bride price! It is reputedly somewhere in Ethiopia; if you look you will probably find lots, for things get copied. I expect the gold covering is missing by now though. It would have been taken off to disguise it, and maybe mislaid. If you find it intact, be careful how you handle it though; it has been suggested that it was capable of developing a static charge, like a Leyden Jar. The Israelites were expressly forbidden to touch it; when one of them did, because the cart was joggled on a rough road, he dropped dead. David was a bit worried by this, and it was sent off for someone to keep an eye on it. Later, a warring tribe stole it, but gave it back because it caused so much trouble.

The measurements are probably quite a lot older than the ark or the coffer though. It seems that there was a previous culture that had a good knowledge of the size and shape of the Earth. When I was young, I visited Stonehenge with a friend who had a car, and was seized with the idea of visiting as many ancient monuments as he could in one day. My first impression on seeing the site was that I would not have put it there! The reason being that there is a more attractive site to the North as you approach it from the road, where it would be on top of a slight rise and show up much better.

We then went on to Silbury Hill. It is said that it was built out of six sided blocks of chalk. The immediate question is why?

One possible reason is that some rocks work better if used in the plane they were dug up; slate for example is much stronger if you use it with the bedding planes horizontal. Some other rocks are more permeable to water in one direction than another. You cannot get six sided blocks sideways without noticing, so that is perhaps an outside possibility. Another reason is that it is very easy to mark out standard six sided blocks with a compass, which could be just a forked sapling, maybe with a couple of bronze ferrules. First decide how big you want them, which is probably related to what you can easily handle. Then you scribe a circle, then step out the radius around the circumference, and you have your six sided block marked out. They fit together very well, like cells in a honeycomb; also, the next row can be stepped one half radius, giving three joints across the centre of the lower block. If you are going to leave voids or chambers in your hill, this is quite a strong way of doing it. Also, if you wrap a bit of grass rope around your six sided block, to protect the edges, it will roll quite well, much better than a square one. There is also less cutting per unit volume than making a square one.

Regarding the siting of Silbury Hill, I would not have put that where it was either. It starts in a valley and is more or less tagged on to the side of and partly cut out of a much better natural hill. Moving it a hundred yards would give a much taller hill for much less work. So obviously these early guys had got things wrong. Or maybe I had. After all, they had built the things and must have spent quite a lot of time planning them.

Thus I was forced to the conclusion that these monuments were built exactly where they were because that was exactly where the builders had wanted them built; they were accurately located, perhaps to a few feet or so, on the surface of the earth, which meant that early man had GPS, which had not been invented either for them or for me at that time.

So how had they measured so accurately? Exactly the same way as we did before GPS, by measuring the angle of stars, including the Sun. The circle is divided into 360 degrees. Don't ask me why, but it is a very convenient number, with lots of divisors which make calculation in your head easy. I spent ages trying to work out how they divided the circle into 360 degrees; it is easy to divide it by six and by five, but to get to one degree you have to divide it by nine also, which I can't do, nor you either if you try.

They knew there were 365 and a quarter days in the year (or very nearly; they knew to skip a leap year every hundred years and again every thousand to keep in step). They added the extra ones as a holiday, 'the five days over the year', and a jolly good idea too. The Romans adopted the Egyptian civil calendar after theirs had gone wrong. Mark Anthony went on about it to Cleopatra, and she said,

'Why don't you go back to the Egyptian civil calendar, dear?'

And they did, or nearly. Later we had to correct it again, leading to riots about eleven lost days. Britain changed a bit later than Europe, and in Arab lands they still have not sorted things out. At Jeddah airport there is a clock with four faces showing four different times: GMT, American Standard, local time zone, and local Arabic (daylight is divided into twelve hours, so the length of the hour varies as to the season.) The calendar was about seven hundred years late and they were never sure when Ramadan was going to start, as it depends on a first new moon, and is several days earlier each year. The very best calendars were done by the Aztecs however, or maybe they got the information from some previous people. They knew the precession down to the last second and could tell you eclipses like reciting a shopping list. They worked things out as a series of interlocking cycles, like cog wheels, as did the Egyptians. There is an astronomical calculator that uses cog wheels in just such a way. It was brought up as a lump of corrosion and concretion from an ancient Greek wreck near Antikithyra, so that is what it is called. Long after it was brought up it was x- rayed, and showed lots of clockwork. The idea of a 'clockwork universe' long predates Newton.

The earth is about 24,000 miles round, so one degree is about 68 miles; degrees are divided into minutes, so very roughly a minute of arc is about a mile. As minutes are divided into seconds, a second of arc is very roughly ninety feet or so. With a very long baseline and repeated measurements of, say, the occulting of the pole star by a cross hair at night, this kind of accuracy could easily be exceeded, the more so as they were not trying to measure where they were, but to locate a particular position and put themselves in it. This could be done by 'hunting': that is, by taking a series of measurements and averaging. Some survey stones called 'omphalos', meaning 'navel', are equivalent to our ordnance survey trig points. In Egypt they had warning notices telling people not to move them. Even if you bumped one accidentally with your cart you had to report it to the authorities, so they were concerned with fractions of an inch.

In the days when sailors used sextants they did just this, taking three measurements to form a triangle or 'cocked hat', and placed great store by it. Actually, it is by no means certain that you will be somewhere inside your triangle, for several statistical reasons I won't go into now.

Latitude is easy, longitude a bit more difficult, as you need a clock. Luckily, the sun, moon and stars form a very accurate clock. They proved unreliable when navigating the Atlantic at mid latitudes because they are obscured on occasion, like just when you are approaching a rocky coast. For the early builders this did not matter; they could wait until it was clear, probably collecting records over many years, or many centuries.

Nowadays, every leisure sailor has GPS and knows exactly where he is at all times: tied up in the Marina, because modern sailors never go to sea anymore.

Exactly why they wanted such great accuracy is a mystery. Perhaps it became a pursuit of excellence, just as we chase smaller and smaller particles nowadays, or want to look further and further out into the universe. It is often said that they made a calendar so as to know when to plant their crops. You do not need an accurate calendar to plant crops however; there are plenty of other signs of spring in the shooting of buds and the lengthening of the days, and you will plant according to the weather plus or minus a week or so, not on a set day. It is also said that it was to do with religion. There is no evidence for this. Religion is more likely to be a decay product of the knowledge, once people had forgotten what it was all about, or maybe it was religionised for the masses, to keep them ignorant but observant of certain dates, or maybe deliberately codified and simplified so the knowledge would persist long after the way to work it out had been forgotten.

At all events, knowledge of the stars and the measurement of the Earth was a major pre-occupation at one time. It seems that they had to know, and were willing to spend a huge amount of time and effort to get things accurate. They knew about eclipses, the planets that you don't need a telescope for, and the precession of the equinoxes, that slow wheel of stars that takes over twenty four thousand years to complete. Maybe they were afraid of something. Perhaps there was an old memory of a time when the stars did not hold their courses, when the Earth was perturbed in its orbit and the poles shifted, and they wanted advance warning if it started to happen again. We are just beginning to start a space-watch, to look out for rocks that might impact the Earth, for we know now that it is not a question of if, but when.

So our civilization might find itself in a situation where it has to move big rocks again, maybe weighing a million tons or so, possibly by four thousand miles (the radius of the Earth), with the problem that it has to be done in space with the rock moving at several thousand miles per minute, and a limited time to do it in.

So how do we do that? Actually, it is quite simple; you do it by pushing, or possibly pulling, not necessarily very hard, but for a long time. We can start in a small way with some experiments on a railway truck. Suppose it is full of coal to make up a total weight of ten tons. Now, if we could drop it over a cliff it would accelerate at a rate of 32 feet per second, every second. Similarly, if we could give it a ten ton push along the track it would accelerate at that rate. Suppose however that you can only give it a push, by hand, of one hundredweight; that is only a two hundredth of its mass, so it accelerates at one two hundredth of 32 ft per sec per sec. (friction excepted) and would take three minutes twenty seconds to get up to speed. The smallest push will move the biggest weight, provided you can wait long enough to see if anything is happening. I once moved sixteen thousand tons of steel by pushing. The occasion was when I was twelve or thirteen, fishing at low water neaps between two cruisers moored on the battleship moorings (two massive buoys over a quarter of a mile apart, later replaced by concrete caissons) in Portsmouth Harbour. One of the cruisers was I think the Bristol, or Birmingham, and the other was city class also, which I think weighed about eight thousand tons. It was a hot August day; the fish weren't biting, so to get out of the sun I chimneyed up between the two boats where the bows came together, and pushed. After a few minutes they slowly came apart, so that I had to get down into the dinghy, after which they equally slowly moved back together!

Actually, although I moved sixteen thousand tons, or thereabouts, there are some caveats. I could probably push at about one tenth of a ton, for I could easily support my own weight on one leg, and I was pushing with two. This gives an apparent ratio of one hundred and sixty thousand to one. However, pushing two boats apart is not the same as pushing one of the same mass from a fixed jetty. To start with, both boats move, and the two hundredweight push is really four hundredweight, as one is pushing in two directions. It is in fact the equivalent of pushing not sixteen thousand tons off a fixed jetty, but four thousand tons. Also, I was only pushing out one end of the boats, so it becomes effectively two thousand tons, or twenty thousand to one. If you then divide by thirty two to get feet per second, then by 60 to get feet per minute, you will find that only a few minutes' pushing will give a measurable effect.

So, supposing that we have to move a few million tons to one side so that it misses the Earth, what do we need for the job?

To start with we need to identify our lump of rock, with speed and trajectory, at the earliest possible moment. Hopefully we will have the gear to do the job on standby, and not have to design something from scratch in a hurry. What we ought to have ready is a rocket to launch a power source, which will fire some of the material off our lump of rock, or ice if it is a comet, using the material of the comet or meteorite as its own reaction mass. This saves having to ship thousands of tons of stuff up there. The power source can be a small simple nuclear power plant, and the 'rocket' or mass driver can be a simple electromechanical catapult, chucking ten tons every few seconds as fast as possible. If it is ice, it is even simpler; we use a high pressure nuclear fired boiler, say at three thousand lbs per square inch, and fire a jet of steam off, maybe diverting a little to melt our ice and pump up into the boiler with a dejector. We could take another route, and vaporize some of our asteroid or comet to fire it off as a plasma jet. This way however we dump an awful lot of energy into the departing ion beam for very little thrust, so large masses at low velocity seem best. What we really need is the biggest push at the lowest cost.

There will be problems, of course. To start with, our lump is likely to be rotating, so we will first have to kill its spin, which will take time. Then we may not have a very good surface to sit our device on; perhaps screw legs might be in order to bolt it down to a melting iceberg or shifting gravel bank. Then we have to arrange to dig up a supply of its bulk to use as our rocket fuel. A dredge, dragline or Archimedean screw might do the trick; perhaps the whole works as a sort of belt and braces idea, for we may not have time to go back to the drawing board. Once we have the device up and pushing, then all we have to do is sit back and take measurements, maybe for months or a year or so if we have got our spotting technique together. It might be a good idea not just to deflect it, but to deflect it in such a way that it is not going to come back again in a hurry; maybe we should tip it into Mars, to bring the mass up a bit and give it some atmosphere, while we are about it. It would make a lovely firework display and concentrate people's minds on the job in hand, as they will have forgotten about the comet that hit Jupiter, which was rather a long way away in any case. We could have a few different thrusters of different sizes, parked in orbit, ready for use, just in case. True, the chances are very low, but the destruction will be massive if a stray rock hits. Also, we may have got the statistics wrong, for they may go around in groups, making a nonsense of our 'one every ten thousand years' kind of reasoning.

Once we have got good at moving rocks, we can start another stone age. We could move Mars nearer the sun to warm it up a bit, by crashing rocks into it against its orbital direction at twenty odd thousand miles per second. We could steal a moon from one of the outer planets; they have got plenty, and a big icy one could be crashed into the surface to give an ocean or two. If we liked, we could explore the Oort cloud to see if could find any useful hydrocarbons to make plastics from, so we could set up greenhouses to get plants going. When sea level rise get out of hand on Earth, we could move the moon out a bit, to reduce the tides, or conversely move it in a bit if tidal power is more important.

All in all, we could get everything in a terrible mess and tangle, with vested interests saying everything was all right, while they all sued each other. The solar power generators would be suing the orbiting sunshade company; Esquimos would be complaining that their new modern houses, provided to replace igloos, were sinking as the permafrost melted. Subsidized American peanuts from Mars would be dumped in orbit on being refused entry. Realtors would be selling Venus on the promise that they would get it sorted out once they had got enough money in. Somebody would accidentally drop a high carbon moon into the sun, unfortunately catalyzing the formation of helium a little bit and speeding up the burn. While everybody's attention was diverted someone would have put advertising all over the back of the moon, and then given it a spin.

How do I know all this will happen? Because we have done this kind of thing before; we are good at it! Man is not primarily a tool making animal, but a rubbish making one.

A VERY DEEP HOLE

It is strange to think that although we can look up and see galaxies that are thirteen billion light years away, and explore mars by robot, we know very little about what lies a few miles under our feet.

Mining can take us down a few miles, and drilling perhaps three or four times as far, but after that we are reliant on sound waves. These can tell us a lot about structure, but not so much about content. For example, it seems that the central core has a density of about ten to twelve as opposed to seven-point something for iron. It has been suggested that the extreme pressure makes the iron denser; maybe, but I think it is more likely that we have something inherently denser there. At the centre the density is about thirteen and a half. Any heavy atoms are likely to sink to a lower level; there will in effect be a sort of 'atmosphere' within the molten iron, just as there is a very slight preponderance of heavy hydrogen at the bottom of deep oceans. Gold, platinum, palladium, iridium and osmium, to mention just a few, are all very heavy metals with atoms of low mobility that might well settle out towards the centre. Once there, convection would be very slow, due to the much-reduced gravitational field, so they are likely to stay, along with a host of other heavy stuff mixed up with the iron. There is an outside chance we might be able to detect denser atoms by examining neutrino transits, but probably the fine-tuning of an experiment might be difficult. Thus it would be interesting if we could go a little deeper, perhaps sending instruments rather than attending personally.

The twin problems are heat and pressure. Deep mines tend to close up without constant maintenance, for rock is to a degree plastic and the higher the pressure the faster it flows. Eventually as you go deeper the pressure will exceed the strength of any material you may use to keep the walls apart, although it will become hopelessly uneconomic long before that. Also, it will become too hot for comfort, as the temperature rises a few degrees every thousand feet as one approaches the molten core of the earth. Somebody once suggested that we could send a slug of molten iron down, heated by radioactive decay, a sort of China Syndrome as it is called by nuclear engineers. Well, maybe.

Drilling is faster and cheaper, but again there are limits. The drill has to be turned and there is friction all down the well. Similar problems with pressure occur, although pouring in a heavy substance like baryta slurry can overcome this, until the temperature rise starts to cook it and boil off the liquid element. The Mohole project was supposed to reach twenty miles, but I am not sure if they actually got that far. At all events twenty miles or less is probably the practical limit for current drilling methods.

So how could we go deeper? Strangely, the oldest of all techniques, pile driving, may come to our aid. The twin problems of heat and pressure may be made to work in our favour. The deeper you go, the more plastic the rock and the more easily it can be penetrated.

Supposing we could drill a hole to twenty miles, or perhaps some lesser figure, we could then lower a steel pile down that hole. Steel has a density of 7.7 and a cubic foot weighs 477 lbs. There are 5280 feet in a mile, thus a steel pile of a foot cross section twenty miles long would weigh 22,487 tons and exert that pressure per square foot at the bottom of our well. Interestingly, you would not be able to pull it up again as that weight exceeds the tensile strength of steel! We could increase the pressure by another few thousand tons by putting a weight on top. We could try banging it in also. I once had to hammer a length of two-inch water pipe a yard or so into the ground, for a lamp pole. I was not getting very far with my eight-pound sledgehammer, so I tied a fifty-gallon water tank on it and filled it from the garden hose. The combination of my banging, plus a static load of five hundred pounds did the trick. However, we might not need to bang it. At around two hundred tons per square inch our hot rock would start to flow out of the way of the pile; it would be rather like putting your elbow in your porridge.

If we did have to bang it, it would take the shock wave from our piledriver about 25 seconds to get to the bottom of the pile. The mass at the top would act as a reflector for the return wave, so we could get a free ride by timing our next hit to coincide with its arrival. As we got deeper the weight of the pile would certainly start to carry it down of its own accord; there is plenty of energy to power it. Drop a cubic foot of iron twenty miles and you have over 25 horsepower hours of energy, a lot better than the average hydroelectric plant! There would be some upthrust of course; the density of the rock might be about 2.7, leaving us with a net effective density of 5, but that is plenty.

Eventually it would start to melt, but we would be quite a long way down by then. Finally we would have iron penetrating iron, with no density differential; it would still keep on going though, due to the weight of all that pile above the iron drill. We could use tungsten, which is about three times as dense as iron, and nearly as heavy as gold. Many gold bars in bank vaults are reputed to have tungsten cores. Who wants to know? There is lots in Cornwall if you want to have a go, but it is hard rock mining, and the seams are usually thin. Smelting is costly too; you have to use an electric arc furnace to get a solid product. The oxide can be reduced with carbon, but a powder is formed which has to be heated, compressed and hammered to get a solid product. The melting point is 3370 degrees, the second highest of all after rhodium. You might cut costs in mining in Cornwall by way of by-products of the mine, such as tin and a few other metals.

Our problem might be that it would go too fast, so we would have to devise a magazine loading device, where sections were automatically screwed on, and slid up a curved ramp to the vertical, to disappear down the hole like spaghetti.

Just putting a pile into the ground that we cannot get out again, and which may sink out of sight, is not very useful, so we would have to design a robust instrument package to go with it. As the end would abrade or melt we could put in serial packages, partly for redundancy as they burnt out, and partly to take measurements at different depths all the way down. We could cool the tip if we liked, pumping down a heavy inert liquid, perhaps with graphite dust as a lubricant, and this could be a power source for our instrument package as it evaporated, or we could use high temperature insulated wire, until the copper melted. A light pipe might be better, as it can be made with a higher melting point than copper. Silicon or silicon oxide might be better than soda glass. We could use sound waves to get signals back if a wire proved unsatisfactory, as sound travels well in iron. There is a lot a robot package could achieve; temperature, pressure, chemical signature, gas content, speed of sound, reflective layering, birthday greetings, and so on, but I will leave the design to the specialists.

If we could drill into a subduction zone, there might be a case for sending nuclear waste down, particularly if we had a continuously sinking pile where we could add package after package for the price of one hole. It would be several million years before the material surfaced again, so long as we did not start a volcano off with the extra heat generated by radioactive decay.

If anybody does have a go, I would love to be invited along. The inventor always seems to get left out, except for a footnote after you are dead. I try to tell large organisations. 'Never steal inventions. Steal inventors!', but they don't listen.

I could make useful suggestions to the engineers, encourage people generally, and get in the way. I suppose they have a point.

ABOUT TIME

'Now, at this present moment in time...' Do you cringe when people say that? It belongs with the foot pedal, and the manually operated hand handle. Why can't they just say 'now', or 'at present'? It is a repetitive, unnecessary, redundant superfluity. I know all about it because I do the same thing myself, and then have to try to take it out again when I do all the typos, and the revision, and the eventual rewriting when that does not work.

However there is a bigger problem with time. Everybody thinks they know about time, until they try to explain it, as one cleric said a few hundred years ago. The problem is that we make too many easy assumptions; we invent something, give it a name and it becomes real to us, a familiar entity that merges into the background and skews our vision, as unjustified assumptions are wont to do.

Terms like past, future, present, and the flow of time trip off the tongue as if we knew what we were talking about. If time flows, where does it come from, and where does it go? Actually it probably does not go anywhere; it is just where you left it, minding its own business and propping up today. It might be better to assume that we are the passengers and time is stationary. We tend to think that time is measured by clocks. After all, if you look in your gas meter, you will see bellows pumping in and out as the gas flows by, or see a little propeller in your water meter. Break the lead seals and open your electricity meter and you will get quite a shock; there is electricity there even if you cannot see it, and it has to be measured by its magnetic field. Take your clock to pieces however, and you will find no time. Our clocks are empty.

Again, what do the terms past, future and present, really mean? If the past is that which is gone and the future yet to be, then the present is merely an interface, a surface with no thickness, so that does not exist either! We tend to think of the past as more real than the future, for the past is fixed, and the future mutable, with potential choices to be made. The converse idea could be equally valid; here the future is immutable, but as it goes through the present it is shredded into different pasts, leading to many roads to the present, although you only travel one of them. However the idea is not attractive, for if we abandon the concept of an optional future we also abandon the concept of free will; a dangerous step, for all the progress we have made has been made by people who really thought they could change things, that they could make a difference. If they all give up then we have stasis, which will almost certainly tip over into decay sooner or later. Or maybe the present does have some kind of thickness; there may the smallest possible unit of time, equivalent to the Planck distance, which is the smallest possible unit of distance. This is very small indeed; one to the minus thirty-five noughts when I last looked, although it was in an old book. Perhaps quantum realities do not part like broken glass, but more like Velcro, at the scale of the very small.

In other times, other people have had a different take on time; there is a passage in the Bible (I will leave you to find it, I don't have my reference to hand just at present), to the effect that 'That which is past, is yet to come, and that which is in the future has already been, and God requires the past.' Quite why God should require the past is not clear; maybe there is a hidden layer of meaning in that 'require' is active rather than passive, a kind of repossession, and the past can be re-worked. Although Aramaic can be translated into English, something can be lost in the process, for it contains 'pools of meaning' where a more diffuse knowledge is presented. There is also an old Jewish saying, that 'to see the future you must walk backwards.'

Also, we are entitled to ask 'whose time?' For my time is not your time if we are moving relative to each other, or if we are in a different gravitational field, say, with you at the top of a building and me at the bottom. Even that small difference is measurable. At high speeds the problem becomes even more acute. Ride a photon, and you have instant elsewhere, for the photon lives in stopped time; it has no time of flight from its internal point of view, we impose that on it from our time. If we could travel at the speed of light we would see the universe expire on the instant, all of time compressed into one sheet, while an external observer would see us motionless, watching our stopped clocks. We would be flat also, for time is a sort of dimension, and we can trade it for a space dimension, have to in fact. Einstein explained it all in his theory of relativity, which I am not going to expound here, as there are many good books on the subject. Don't be put off by thinking it is difficult. The broad concepts are simple and logical, and only the detail and the maths are more complex, well to me anyway. There are some loose ends and paradoxes. Take the 'Twins' paradox for example: if one of a pair of twins takes a space trip on a very fast rocket, on his return he will be younger than his twin who stayed at home. This could be quite useful if you go exploring distant stars, for you will age more slowly and thus be able to travel further than if you kept the same time as at home. On your return you might find your grandchildren had died of old age however, for time will run at its old speed for those who stay behind.

A further twist to the paradox is that both twins could have experienced the same gravitational field, if the rocket can accelerate constantly at one 'g'. This is explained by saying there is a difference between a gravitational field and an accelerational one, which contradicts the principle of equivalence that says they are the same, a not inconsiderable problem as the principle is inbuilt into the theory. One can get round it by saying that one twin is travelling and the other not, but this falls foul of the bit of the theory that says all motion is relative.

I remember first reading about the principle when I was about twelve; I came a bit late to it, but it had not received the general exposure then. I remember thinking "This guy cheats!", for a gravitational field did not seem at all the same thing as an accelerational one to me then. Actually there is a third field, a centrifugal one, or centripetal if you are a purist.

They are all different. A gravitational field is spherical, and decays by the inverse square law if you go outwards, and a bit more sharply if you go inwards. An accelerational field is planar, and does not vary if you go above or below the floor of the lift. A centrifugal field is cylindrical, and increases by the converse square law if you go outwards from the centre of rotation and the inverse if you go in. An accelerational field is the only one that requires a constant supply of energy though, and this is probably the reason why it affects time; it is the same as an increasing mass in the sense that energy and mass are interchangeable. They are the same sort of thing inasmuch as you can use one to negate or reinforce the other. We are not flung off the earth by centrifugal force because gravity is much stronger, but you weigh slightly less at the equator nevertheless. The moon is towed around by the earth, as it is captive, not having sufficient energy to climb the gravity well and go off by itself. As a result of this interaction there are null points, the LaGrange points, ahead and behind in its orbit, where object could be parked and not fall anywhere. The moon is actually moving away slightly, due to a slingshot effect caused by the tides. This will eventually die out as the tides stop the earth turning, so it will never get away unless something else comes along and kicks it out of orbit. The moon has already stopped turning due to its smaller mass and greater tidal effect due to the greater gravitational effect of the earth. Tidal effects don't need water; rock distorts slightly as well, leading to frictional loss of energy, and the Moon would have been molten initially anyway.

We usually assume time is like some kind of road travelled, with the option of side turnings in the future, but not in the past. This possibility of choice is behind Ritter's 'Many Worlds' theory, although the concept existed before he enunciated it. Here all possible worlds exist, but you only experience one of them. As somebody once commented, it is 'light on theory, expensive in universes.' There are no experiments that can be done at present to explore two or more worlds together, although I have touched on one in another article. We might examine identical twins perhaps, but identical twins are not really identical in fine detail. The concept is inherent in quantum theory however, with its 'super positional states', in which two possibilities are thought of as being entangled, only to be expressed as an 'either or' possibility when the state is tripped into one or the other.

We also assume that the past is immutable, for no good reason, except that changing the past leaves one with a feeling of unease; no firm rock to stand on. The concept of the past disappearing entirely does not give the same unease, which is illogical.

It does seem that some people can see into the future, or unknown bits of the past. Psychics are sometimes used by the police to find missing persons, and they are rather hard headed. Despite fakery, some mediums do seem to have access to information they could have no direct knowledge of. Animals may have the same kind of ability. Maybe we once did have a greater ability, which declined with speech, as we have to codify everything to express it. Nowadays we seem to think that everything can be expressed in zeros or ones in our computer binary, shortly to be further compressed to into 'q bits' in our quantum computers. The next step will probably be to realise that nothing really exists at all and that we are figments of our own imaginations. The present level of understanding is rather unsatisfactory.

Oh well. Perhaps time will tell!

SOME LIGHT EXPERIMENTS

It has been said that the Greek philosophers did their work so well that they stalled progress for over three millennia.

In our own scientific age a similar effect may have been produced by great men like Newton and Einstein; such great advances were made that there was need for a pause to assimilate them. Even now, the man in the street is not very aware of the implications of what they actually did. Sometimes those in the scientific fraternity explain this lack of general understanding by saying that their work is too complicated for the average person to understand. Unfortunately the perception may be that it is too abstruse to be worth paying for also, and there has been a steady reduction of funding for 'blue skies' research in the past few decades.

There is a case to answer: often great advances in understanding do take time to become applicable and useful to the man in the street. Newton's work on optics for example seemed to leave nothing else to be explained. If you read comments by scientific thinkers at the beginning of the industrial revolution however, there is a feeling that much of his work 'lacked utility', and it was time for more practical men to solve more pressing problems. It was not until the development of Quantum Theory, more particularly Quantum Chromodynamics and the application of Maxwell's equations, that optics took off again. We now have a rich field regarding entanglement and quantum computing to explore, plus many other effects.

There are still however some questions that can be asked by the layman, and I wish to set out a few here, as a basis for discussion to see if anyone can come up with the answers, either from within or without the mainstream.

Light is the visible form of electromagnetic radiation and it is composed of photons. We tend to think of the photon as a sort of fundamental entity; the smallest discrete quantity in the same way that atoms were at one time considered fundamental. However this maybe a simplification; perhaps it has many parts. For example, it has an electric field and a magnetic one, it can be polarised in several directions; it can give, depending on how you look at it, the impression of being a wave or a particle. It can come in an infinity of wavelengths, has no mass but possesses energy; it has a sort of 'reaction mass equivalent', in that it will exert a force if bounced off objects and it always travels at the same speed. It can also 'sense' the presence of a suitable medium to travel through.

If you set up a right angled prism, with light being reflected at 45 degrees off one face, then 45 off the next, all of the light will be reflected back along a parallel path to the incident ray, losses due to less than perfect transparency excepted. However, if you bring up another prism against one of the faces from which it is being reflected, some of the light will pass into the second prism despite a gap between them. The transmission will increase as you bring the second prism closer until it all passes through, provided you have perfect faces. We can explain this by using quantum electrodynamics, which, in a nutshell, says that virtual photons travelling at infinite speed, which they can do because they are not really there, explore all possible paths and then take the most economic one. Thus the photon has a surprising range of properties, which might indicate that it is not so simple after all.

It is often said that nothing can go faster than the speed of light. This is not quite true. Information has zero mass and zero energy; it is not a 'thing' in the normally accepted sense and there is no reason why it should not travel instantaneously. Also, in talking about the speed of light, we have to ask, 'whose time are we using?' Time slows down for the high velocity observer from the viewpoint of another observer not travelling at that speed. Thus, if you wish to travel to the nearest star at somewhere near the speed of light, your elapsed time by your clock will be less than that of observers at either end.

Many experiments with light use reflection to shorten the compass of an experiment, and the assumption is that it occurs instantaneously. Maybe it does, but it is at least a testable hypothesis. At all events, reflection is not an entirely surface phenomenon; the light penetrates slightly into the reflecting surface, as can be shown by the polarisation of light from a polished magnetic surface.

Most people tend to think in terms of physical objects, and there is something contra-intuitive about the instantaneous stopping and restarting of a photon in the opposite or any other direction. For example, if a tennis ball is 'reflected' off a wall, the ball deforms; at first contact the front of the ball stops, and energy is stored in the rubber as the rest piles in behind it. Then the previous trailing edge starts to reverse direction as the rubber starts to regain its shape, followed by the rest of the ball. There is of course some energy converted to heat, but ignoring this we find that the velocity averaged for the two paths before and after reflection is slower than that of a straight path without reflection.

The reflection of light is somewhat different from that of a tennis ball, but the photons do become entangled to a degree in the reflecting surface; if there were no interaction, there would be no reflection. Thus there is potential for a sort of 'dwell time' while they are being entangled and disentangled, this apart from any slowing due to the short traverse of a different medium. This could be tested by multiple reflection of light between two mirrors. If we could accurately measure the distances of the multiple reflection paths, perhaps using a beam splitter to send half of the light down a straight path, we could recombine the two beams to see if there was any phase difference at the end. We could then subtract any difference caused by transmission before reflection to see what the net result was.

Next I wish to examine the different speeds of light in different substances. The general perception is that light travels at its highest velocity in a vacuum, and slower in other transparent media. The reason for this is that the wavelength of the light interacts with the wavelength of the outer electrons in the substance it is passing through. Recently it has been shown that by a suitable selection of media, light can be slowed very greatly, even stopped temporarily until kick started again by a reference beam. Whether we interpret this as stopped light or the energy of the light stored temporarily is a debatable point, for a stationary photon does not make a lot of sense; however this is the net effect. We can also show negative refraction, and it is possible to send a radio signal through a germanium crystal at an apparently faster than light speed. Whether this is a 'water pipe effect' is debatable. If you turn on your tap, you get water out in a split second, although the reservoir may be hundreds of miles away. The water you get out your end is however not the water that flows into the other end, so the speed of water in the pipe is not the distance over the time taken, but much slower than this. However, it was realised over a century ago that the greatest refractive effect was obtained when the period of oscillation of the electrons in a substance was close to that of the light transmitted through it. If a wavelength slightly shorter than that of the natural vibration was chosen, it should propagate at greater than the speed of light.

Another interpretation could be that the speed of light is the same in any medium, but that in some it takes a longer path. For example, two runners, one running along a field edge and the other through a plantation of young trees, might actually travel at the same speed, but the one that has to dodge round the trees will appear to be slower due to the longer path taken.

This is possibly a testable hypothesis, for we can test the frequency or wavelength of light in different media: Many substances are photosensitive, but only to certain frequencies; quite often the cut-off point is quite sharp. For example, you can take a picture with your camera in a thousandth of a second. You can then take the film into your darkroom and expose it for a million times as long; at perhaps ten times the effective aperture, while developing it under a dim red light.

Thus we could have a photosensitive substance, soluble in two or more liquids of different refractive index, which could be exposed to light just at the edge of its cut off frequency, perhaps using a tuneable laser. If we find there is a difference in sensitivity, we have different speeds for light in the different liquids, if not, we can argue that the speed is constant, but the effective paths longer.

Next we can examine the speed of light in a moving medium. For example, we could note the time taken for light to traverse an evacuated tube. Then we could fill the tube with a gas or liquid, and note the time again. We could then start the fluid moving; if we used hydrogen, we could get it flowing at many thousands of feet per second. While this is but a small fraction of the speed of light, we could use a long tube and a short wavelength, and also run the experiment in two directions, with the light travelling against the flow in one, and with it in the other. This would easily bring the experiment well within any experimental error. We could then find out if the light is slowed down, rather as a boat is slowed down moving against the stream, or if the wavelength is altered. If we could move a low-pressure gas faster than traversing it slows light down, we could move light faster than light; a kind of moving escalator effect.

We could also, instead of moving the gas through the tube, move the whole tube, with the gas contained in it. The effect should be the same; but what then if we move a tube with a vacuum in it? Most people have no problem in envisioning a movement in relation to a vacuum, but what of a vacuum moving in relation to an observer? The idea seems nonsensical, but why does it seem so? It may be that we subconsciously have substituted the vacuum for the ether. We tend to think of empty space as being filled by the vacuum, which is another way of saying it is 'full of nothing', which is a nonsense statement. One cannot have relative motion in respect of nothing.

We are used to the idea that the speed of light is a constant; we could extend this to say that in any frame of reference the vacuum is static regarding that frame. This solves that particular problem, but opens up another. The vacuum is not really empty at the level of the very small, but filled with virtual particles. If we have two reference systems, say two gas clouds interpenetrating at high velocity, we open up the possibility of each set of virtual particles interacting with each other. While they may each have zero net energy if considered separately, they do not if compared to each other, by virtue of their relative motion.

Although this concept of zero relative velocity with regard to the vacuum is convenient in some instances, we have a problem when we consider a change in velocity. For example, we can be happy with the idea of an observer at rest in relation to his local vacuum, but how far does local extend? Presumably there is no cut off point, so we are still at rest in relation to quantum particles halfway to Proxima Centauri. If we then change our velocity we have to consider such alien subjects as the 'rate of propagation of stasis'. We could propose that the rate of change occurs at light velocity, rather like gravity, but this is just an assumption, without any real basis.

Another experiment could be done with a rotating glass disc; this time the movement of the medium will be at right angles to the light beam. We first set up the experiment to determine the speed and direction of a light beam shining at right angles through a stationary glass disc. Having got our reference, we then rotate the disc as fast as possible. While the light is in the glass, the glass is moving at right angles to the direction of travel of the light. Is the light carried sideways while it is entangled with the glass electrons? Is the light beam refracted, for it is in one sense entering the glass at slightly less than a right angle? If so, what is the net speed of light in the glass, allowing for the accepted change of velocity for stationary glass? Also, what of the frequency and direction when it leaves the glass? For the component of velocity will remain, rather in the same way as someone stepping off a moving bus takes the velocity of the bus with them.

There are some even more interesting experiments that could be done with rotation. I will first of all deal with an experiment in a classical sense, to set the scene, so to speak, then substitute light for real physical objects.

We use a large centrifuge; one of those hand powered merry go-rounds to be found in a children's playground would do. We need two observers, one rotating with the centrifuge and one standing to one side, and an event to observe.

For the event we will use two billiard balls, one red, one blue, and a short length of tube they will fit into, containing a spring and latch mechanism. The two billiard balls are pushed into the tube, compressing the spring, and the latch prevents them flying out again until tripped.

For our first experiment, we have the centrifuge not rotating. The internal observer is at half radius holding the tube tangential to the radius, and the latch is tripped. Both balls are ejected, following identical and opposite trajectories, both striking the rim of the centrifuge at the same time. Both observers compare their observations and find that they agree.

Next, we set the centrifuge rotating, the latch is tripped, and both observers take notes. The external observer sees the two balls fly apart as before, at a tangent to the half radius; they both follow straight-line paths. However, one (the blue one) is travelling at its previous speed, plus the speed of the centrifuge, while the other, retrograde to the spin, travels at its previous speed minus the speed of the centrifuge. They do not hit the rim at the same time.

The internal observer makes a different set of observations. The balls fly apart as before, with identical velocity so far as he is concerned, but they then follow different paths. It appears that there is a kind of gravitational force attracting the blue ball to the rim; it follows a curving trajectory and hits the rim before the red one, which seems not to experience the same amount of force. Puzzled by this, he does a few experiments. Taking one of the balls and a spring balance, he measures its weight at the centre of the spinning centrifuge, and finds it as normal. At half radius it weighs more, and at the rim twice as much more, there being an extra component of gravity horizontal to the to the Earth's field. He then runs round the centrifuge, first in its direction of travel, finding the field increases, then against the direction of travel, finding that it decreases to zero until he is effectively stationary with regard to the outside world, then increases again. Once released however, it carries with it that acceleration pertaining at its release point.

The two observers compare notes, finding that although they have different observations, they can agree where and when the balls hit the rim. By assuming a pseudo gravitational field for the rotating observer and normal Newtonian mechanics for the static observer, they can bring together the two apparently different observations. To make the experiment more exact they arrange for slits and upstands in the rim, so they can see what effect different speeds of rotation have; at one speed a ball exits through a slot, and at another it is reflected by an upstand. Their experiments always agree.

For the next experiment, if you haven't guessed, they use a laser and a beam splitter instead of the ball and latch mechanism. The light pulse is tripped by connection of two contacts, one rotating, the other not.

The external observer does not note any difference in speed between the two light pulses, but sees that the one travelling with the direction of spin is blue-shifted, while the other is red-shifted. As they are travelling at the same speed, they either exit or do not exit at the same time, the path length being the same.

The internal observer also notes no difference in speed for the two light pulses, but does not see any frequency shift either. The pseudo gravitational field also affects light; the pulse travelling with the spin is bent more sharply than the other, as was the blue ball in the previous experiment, while the other is not as much affected. Consequently the apparent path lengths are not the same, and, at certain speeds of rotation one or the other of the light beams can leave the centrifuge.

On comparing notes they cannot agree; they have both experienced different realities, having two different world-views. Calling in adjudicators does not help, for the impartial observers must come from one or other of these two worlds. We could argue that relativistic effect will bring the results together, but this will not serve completely. The diversion of the light from a straight path for the internal observer is dependent on the speed of rotation, while the escape or not of the light is dependent on the slit spacing. In other words there are two independent variables.

Thus we have a paradox, which perhaps could be resolved by experiment. This looks a bit like one of those 'either or' experiments one gets with the collapse of a probability state, but we get both results at the same time.

There are doubtless other experiments that could be done with light, and much is being done. It would be interesting to hear from others with suggested experiments, or proposed solutions; after all, the layman is in part paying for the work, and there is no reason why he should not be allowed to ask questions.

SECTION 6

DARK MATTER

This section is also about the Universe and all that. If you are a nuts and bolts man you might think it has not a lot to do with you, however, most people of a creative turn of mind are not just interested in better mousetraps, but all manner of things, useful or otherwise, so I make no apology.

In reading about dark matter, there are two statements advanced as evidence I have found that I do not agree with. The first is that there must be dark matter within the galaxy to account for it spinning more as a solid disc, unlike the rotation of the planets round the sun. The other is that there must be considerable mass outside of the visible galaxy, again to account for its spin rate.

Firstly, I would not expect a disc of stars to rotate in a similar way to the solar system, because about 98% of the mass of the solar system is in the sun. If one observes the orbital path of a planet, and then doubles this, one would expect the gravitational attraction to be one quarter at this greater distance, with a consequently more leisurely period.

However, most of the mass of the galaxy is within the disk of stars. If one takes an orbit within this disc, and then doubles it, it is true that the original mass now exerts one quarter of the gravitational field, but there is a much greater mass in between this and the new orbit. If the star field is evenly distributed, at whatever distance the field will be the same, it being scale invariant. This would lead to the rotation being like that of a single sheet of stars. In actuality, there is a central black hole, and there is a bulge at the centre, so the resultant would be a hybrid between the two extremes, with the centre rotating a bit, but not a lot, faster than the edges, proportionate to the distribution of mass.

For the second case, I do not believe that extra matter, invisible outside the galaxy as a kind of halo, will make any difference to the rotation rate. For example, if we could dig a tunnel to the centre of

293

the moon, we would find the gravitational field diminishes with depth, until at the centre it would be net effective zero, all attractions cancelling out. If we then start to dig a spherical cavity at the centre of the moon, spreading the excavated material to level out the surface, anywhere within this cavity the net field will still be zero, the reason being that all attractions cancel out.

What is true for a sphere is also true for a disc; only that which is below you attracts, the rest, if evenly distributed, cancels out. This is also true for the universe as a whole, again, if evenly distributed. Thus the presence or otherwise of extra material outside the disc will make no difference to the gravitational field and spin rate, for anything below it (nearer to centre), within the disc. However, there may well be a mismatch between the spin rate and the observed mass, so there is still missing mass, but a 'flat' spin distribution, or added mass outside the observed field does not add to the case.

Another problem for high rates of spin is that there is a potential for mass transfer by virtue of the energy transferred. If material falls inward due to gravity, it gains kinetic energy, while the attracting mass uses up gravitational potential energy and is to that extent 'run down'. Given that energy has mass, the centre will diminish in mass and the circulating ring increase.

CAN SPACE MOVE?

We are familiar with the Newtonian concept of a moving body traversing a fixed space. What however of the reciprocal concept of a fixed body with space flowing past it? Is this concept meaningful or merely a restatement of the effect in different terms? If space is to move, we must ask, in relation to what? Objects, obviously, but it could also move in relation to other regions of space, for at the microfine level space is not empty, but contains virtual implicate quantum pairs, while at the macro level it contains galaxies.

For example, we speak of the expansion of the universe, not in the sense of it expanding into some pre-existing space, but of new space being created in the expansion. To define a distance, we need two end points. At the edge of the universe, we only have one; we can say that in trying to define a distance we are going from one fixed point to nowhere, thus there is no space to expand into. There are two possible modes. Either some kind of wave front lays down new space, rather in the manner of a snail laying a trail, where the length expands but only by accretion at one end, or by expansion of the bulk, rather like dough rising, where expansion occurs evenly throughout, leading to a greater velocity the further one is from the centre. This latter is the more commonly accepted proposition, although both could co-exist. It leads to some novel conclusions. For example the expansion of space implies the continuous creation of energy. If two galaxies start with little or no relative velocity, expansion will progressively accelerate them apart, leading to an increase in their kinetic energy relative to each other.

As an illustration, if we had a long enough fishing line we could cast a weight from one galaxy to another; as the line reeled out we could make it drive a dynamo. With a ten-pound strain on the line and a recession velocity of a mere fifty five million feet per second, we could have one hundred thousand horsepower from one fishing line! And galaxies weigh a bit more than ten pounds. Also, since their joint gravitational fields will mutually attract them, they will expand apart more slowly than the space in which they are embedded, thus leading to the concept of a flow of space around or through them. Ultimately they will gain a recession speed greater than that of light, and drop out of sight of each other, although there might be the possibility of 'over the horizon' signalling, rather as a relay of optical semaphore stations can pass a message around the curvature of the earth.

Alternatively, to preserve the first law, we could propose that no new energy is created, but the energy is provided by some hidden store in space itself. The process then becomes part of the trend to entropy. Space does seem to contain a lot of energy, although the miss-match between that observed and that calculated is the biggest miss in science. Maybe we are observing some kind of net effect rather than the whole. In this scenario of expansion, the implicate quantum pairs that go to make space, to keep it apart so to speak, will diminish in density as space expands. After all, they exert a pressure, so they should force each other apart if space is expanding; maybe they are responsible for 'dark energy.' We could equally argue the alternative however, and say that if we could harness the energy of space we would in fact run it down, and things would become closer together.

We don't really seem to have a very good handle of what exactly empty space is. If we say there is nothing between two objects, then they are touching. However we do not observe this; they can be any distance apart, which can be shown by a ruler of some sort between two objects, and by the angle subtended by two, to a third.

This presents a problem for the empty set, the building block of set theory. It is now full of an infinitude of nothings, all different, rather like Berkeley's 'ghosts of departed quantities'. We have quantised nothing. We could say that it all adds up to nothing, but if it does not add up, it can't be subtracted either, and renormalisation won't work, merely cycling back to the start point.

We also have the problem of how things move through space; do they carry their own space along with them, or does space flow through them, rather as a flock of birds can pass through the air without dragging the bulk of it along with them? If we assume that the expansion of space implies movement of galaxies, then presumably it also implies the movement of quantum pairs embedded in the moving space. We could have a collision problem, for if two volumes of moving space collide, perhaps in a multiverse, we have a mega-collision of quantum pairs, where the net energy is no longer self-cancelling. A good way to create a universe if you are looking for one.

We can modify Einstein's train experiment to examine relative motion between different quantum pairs. Suppose an experimenter sets up an experiment to demonstrate quantum pairs on a railway platform, while another passes by in a train with an identical experiment. Does each observe their own set of quantum pairs, not in movement in relation to themselves? If so they seem to be rather subjective items, dwelling more in the mind than in an external reality. We have a parallel in the speed of light; each observer has his own personal speed of light, which happens to be the same as anyone else's, irrespective of his or her relative movement. Perhaps everyone has his or her own personal speed for quantum pairs, which happens to be zero irrespective of relative movement between different observers.

We have a problem if we bring the experiments closer and closer together however, until they overlap. What if half of a pair from experiment 'a' negates with its opposite partner from experiment 'b'? Presumably the surplus energy will be supplied from the train, rather as a black hole may evaporate. Alternatively we have pulled energy from empty space, and black holes don't evaporate after all, merely tearing the fabric of space apart, in which case the infalling half pair has positive energy and the hole will get bigger.

There is a further dichotomy to be found if we measure the distance of far galaxies by the time light takes to reach us, as shown by the red shift. If space is expanding, the light will take longer to reach us, for it will be travelling against a headwind of expanding space, not so much going slower, as having a further distance to travel. Alternatively, if the expansion of space is in some way caused by the expansion of the Higgs field, which drags matter along with it, the light will not be impeded by this, for light does not have mass.

There are other interpretations to be made, and other experiments, each with conceptual paradoxes attached. So, can space be moved?

Watch this space!

DOES THE UNIVERSE ROTATE?

Everything seems to spin, from atoms to galaxies. One might wonder whether the universe spins also. We could derive a 'spin start' model for the expansion of the universe; this would nicely explain the origin of particle spin also, as there would be a fossil spin imparted at the outset.

This leads to a problem; spin in relation to what? The concept of a spinning universe with no external reference may not be meaningful. There is also the problem of reaction mass, for to spin something, one needs an equal momentum in the opposite direction. We could propose an equal and opposite universe, but there is no evidence for this.

Newton wondered if the water, heaping up towards the edges of a spun bucket, could in some way 'sense' the rest of the universe it was spinning in relation to. Nowadays we are happy with the concept of a force being generated by the continuing change in direction as the water rotates, although actually this does not really explain inertia. This could be used to see if the universe is spinning, for if so, it is likely to be disc shaped, as in a spinning galaxy. Also, galactic spins would tend to be aligned, rather as planets formed from a condensation disc around a star all spin in the same direction. It would be also easier to spin things in one direction than the other, if only very slightly. There is no evidence for this, galactic spins seeming random, and the universe does not seem to be denser along a plane of purported spin.

Probably the reaction mass for spin is internalised within the one universe, being a relic of turbulence in the beginning. Spin is not conserved however, so we could propose some of the spinning galaxies have been precessed to spin in the other direction, and the rest randomised, but we would need a mechanism.

We could examine spin in more than one dimension, but there are problems with resultants. If we take a sphere we can spin it in one plane. If we try to spin it in two planes at once, then we merely get a resultant at forty-five degrees. However, if we spin a sphere and then fly round it at right angles to its spin, our moving ground zero does not form a resultant at forty-five degrees, but traces a figure eight one side of the sphere only. Quite useful if you want to send a synchronous satellite up to cover one side of the earth only. Adding more dimensions could create other patterns perhaps, but I can't visualise them in my head. It may be that some combination of multi-dimensional spin could give 'anti spin', where space spins but things within it do not, resulting in a negative centrifugal force, or gravity. We could add in a Tippe-Top reversal, where the poles change over without the spin reversing, but again, we need to propose a mechanism for this.

A METHOD FOR DETERMINING THE LATERAL VELOCITY OF DISTANT STARS AND GALAXIES

When viewed through a telescope, distant cosmic objects show a red shift for the light arriving from them. Assuming the effect to be linear, this can tell us their velocity and hence distance. However, it will tell us nothing about the lateral component of velocity at right angles to the line of sight. For example, two objects might have the same velocity, one along the line of sight and another at right angles to it. We can determine the velocity and distance for the one showing a red shift, but nothing about the other. In practice, most objects will have a direction of travel that can be broken down into the two components, the lateral diminishing as a fraction of the whole with distance, but still keeping the same value.

There are two potential methods for finding this lateral component. One is to take a bearing from equidistant points on the earth's orbit to find the difference in the angle subtended. Unfortunately the earth's orbit at about 180 million miles is far too short for all but the nearer stars in our own galaxy. The other, to take a picture on a known bearing, and then compare it with a later picture from the same point in orbit, will lead to a very long wait indeed before there is any measurable divergence.

There are however two other methods that could be tried to find lateral velocity. One is the time of transit for the light across a measured distance. For example, supposing we have a lift with two sheets of paper connected to a ballistic chronograph, and fire a catapult pellet through them at an angle to the normal such as to give a Pythagorean triangle of 5:4:3 feet at a speed of 100 fps. If the hypotenuse of a Pythagorean triangle thus formed is five feet, then we will find that the time of travel is one twentieth of a second.

The lift is then set in motion at 60 fps, so that the back screen is in position directly opposite the entry point by the time the pellet arrives. This will look, to an observer in the lift, as though the pellet has been refracted in the same way as a block of glass will refract a beam of light. If so, we should expect the pellet to have travelled more slowly, and in fact this is the case; it will have travelled four feet in one twentieth of a second at a speed of 80 fps to the eyes of an observer in the lift. Of course it has not really travelled more slowly. The effect is caused by moving the goalposts during the time of travel, but this is the observed effect.

This would be exactly the same for a beam of light. Its speed of course would be unchanged, but the effect of moving the second measuring point relative to the first during the time of transit would shorten the distance travelled, leading to an apparent slowing for the internal observer. It is possible to time light pulses very accurately, so this apparent slowing could be measured and from this the lateral component of motion worked out.

Now, we are unlikely to be able to move a lift at some reasonable fraction of the speed of light to perform the experiment, but we don't have to. Motion is relative, so the lateral motion of the star or galaxy under examination will do; it does not matter which is moving so long as there is relative motion. One advantage in this method is that we do not need to know the distance of the star; the result will be invariant with regard to distance as only the lateral component is measured.

Another method would be to measure the angle of apparent refraction. There is a difficulty here in that it is always with us, in the water of the eye, the lens of the telescope and to a lesser extent in the atmosphere, particularly at low angles.

We can however avoid this problem by placing a true refracting medium in the way of the beam of light. To do this we could have a water-filled tube with transparent ends that are parallel to each other. Float glass would probably do for an initial experiment, while the length of the tube could be that length convenient to the rest of the telescope and the problems of mounting and transport. The longer the better for the finest resolution, but initially a moderately good amateur scope and a bit of plastic drainpipe would do. We could use a liquid with a higher refractive index than water, but as water is readily available, can be dumped for transport and is effectively free, it might be simpler and cheaper to have a slightly longer tube to gain the same effect.

To commence the experiment, the tube should be set up and adjusted to give an image of a static light source; a pinhole at a few metres might do. The light should enter at the normal; that is, at right angles to the centre of the telescope aperture.

The focus is then shifted to a source with a known lateral movement. A pair of colliding galaxies would do, or the edge of a spiral galaxy. An image is then taken, the tube removed, another image taken and the two compared. Any refraction caused by lateral motion would then be apparent and the speed could be calculated.

Thus we might be able to gain a more accurate map of the universe, or at least where everything was at the time the light was emitted. With both the longitudinal and lateral components of motion we could calculate where things are now, and will be in the future.

A JOURNEY TOWARDS THE CENTRE OF THE EARTH

The Da Vinci Institute has formulated eight grand challenges for the future. I wish to examine the first of these: sending a probe to the centre of the earth so as to transmit information back to the surface.

I have entitled my article a journey towards the centre of the earth, rather than to it, for I believe the work will proceed in stages, possibly over a century or more, and may not be ultimately achievable in its entirety. There are two main problems to be overcome; that of temperature, and that of pressure.

Dealing with pressure first, we have a problem inasmuch that any hollow structure would be crushed. The reason for this is that any material would flow at the pressures pertaining, even though it might appear to be a solid. Even in deep mines of only a few miles depth, the walls, floor and ceiling of a tunnel may start to move inwards by plastic collapse, and rock may spall off as the pressure is released on the cut surface. There is an upper limit to the resistance to flow of any material, principally due to the fact that atoms are bound by the forces generated by the electrons in the outer orbit of the material in question. While they are well able to withstand the crushing forces, i.e. those tending to compress them towards the nucleus, at the centre of the earth, with a pressure of several tens of thousands of tons per square inch they are not nearly as able to withstand sideways movement or slippage, and would very easily take up a new configuration that equalises the pressure all around, so any probe will have to be solid state. This does not present a great problem, for we are used to solid state sensors now. Admittedly we might have to use different solids, light-earth metallic oxides such as aluminium or calcium oxide rather than plastic, and perhaps conductors such as platinum or tungsten, able to resist high temperatures without melting at the pressures pertaining.

Thus there is some hope of being able to devise a probe able to resist pressure after all; the centre of the earth is resisting pressure, so a probe of similar materials ought to be able to do so also.

The next problem is that of temperature. The temperature is about four thousand degrees Centigrade, and there is no known element that will not melt at that temperature. Some metallic oxides have a higher melting point than the metal however; for example magnesium melts at 651 degrees C but its oxide is used as a refractory in furnace linings, so we might just be in with a chance, particularly at high pressure. Unfortunately, high temperatures lead to dissociation if the thermal energy exceeds the binding energy, and our oxide turns back to a vapour of the metal and oxygen. Luckily, the pressure comes to our aid here. Melting is generally accompanied by expansion, and pressure inhibits expansion, thus raising the melting point. The central core of the earth is almost certainly a solid, as shown by its ability to transmit shear waves. A liquid cannot do this because it is too mobile, not allowing any lateral forces to build up and produce shear, or sideways slippage. Also, as the centre appears to be a bar magnet, it is most likely crystalline and these crystals might be very large, as they have had such a long time to form. The density at around ten grams per cubic centimetre is too high for ordinary iron though, which is 7.88. If it is all iron, which I doubt, then there is probably a new crystal structure in addition to the stellitic and austenitic that we use in engineering. Carbon forms graphite with a density of 2.2, but under great pressure will form diamond at 3.514. The interesting thing is that it does not revert when the pressure is taken off. Many elements and compounds will probably form new crystal structures under pressure; some of these might also not revert, thus giving rise to a whole new set of materials. Ideal for lining the bore, except that we have to get there first to mine the stuff!

There is still the problem of melting for our probe on the way down however; this presents a challenge. After the crust of a few miles, there is the mantle, a quasi-solid but capable of plastic flow, yielding to a sticky layer of viscous rock, which gives way to molten iron, which is about as liquid as water and nearly as good a conductor. Cooling is certainly possible; all we have to do is to pump down a cooling fluid to keep the probe and shaft down which it is descending at a temperature below its melting point. Water could be used, but we would have to maintain a high top pressure because it is so light. Also, red-hot steam will oxidise iron. The simplest coolant might be liquid iron; if we use tungsten, with a higher melting point for the probe, iron is ready to hand and has the same density as the material it is penetrating, so would no present a buoyancy problem. Lead metal might be better as long as we could prevent it boiling in the upper reaches, as this is nearly twice as dense as iron and would add weight to the probe and drill train, forcing it down.

However, although we could cool our shaft for a certain distance, as it gets longer the problems increase. There is a limit to the flow of coolant we could provide, while the amount of heat it has to carry away increases with distance. Thermal insulation is a possibility, but again, the deeper we go the more insulation we need, which means the bore gets thicker and more expensive. Before long we would be using several thousand tons of insulation per mile, and we have four thousand miles to go. Total loss coolant, ideally perfusing out from the boundary layer, would about double the distance, as we would not have to provide return flow, but again, there is a limit, although we could go a long way with lead. We could tip it in as lead shot to gain advantage from the heat absorbed in its melting, adjusting the column height, and thus the pressure, so that it would flow out at the bottom, acting as a lubricant.

If it is impossible to get a bore to the centre, we could try to drop a torpedo. Not quite the same as the powered version favoured by submarines, but a long slender weighted object that would fall freely through the molten iron, and slowly through sticky rock, sending back information until it was overcome by pressure or temperature, rather like a Venus probe.

We still have to get through the crust and mantle however before we get to the liquid iron. The problem with drilling is that the longer the bore gets the greater the friction. While this can be reduced with a lubricant it is still cumulative, and we will hit a limit, particularly when we get to plastic rock, which is sticky and liable to grip the drill tightly. A tip powered drill, rather than one turned from the top might be in order. This could be powered by the coolant flow. The drilling record is about twelve miles, set up by a Russian team.

A very old technology can come to our aid as we get deeper; that of the piledriver. Having got as far as we can by drilling, we can start to bang our drill in. This will be greatly aided by the fact that we will have a heavy tungsten end section, which does not melt until 3380 degrees C, and has a density of 19.3, the same as gold. It might even get cheaper as we use less in light bulbs. The rest of the drill could be steel, which is a lot heavier than the rock it is going through. We could also impose a load of a few hundred tons on top as well to start with. The pressure wave going down from the blow would travel at about three miles per second, so we could have several waves in train at the same time. By using a weight on the top we can get an extra boost out of the return wave reflection as well, for a pile being driven can tend to bounce with a reflected wave travelling back up the pile. This can cause the top to spall off if not controlled, so a weight on top is a good idea even in normal driving.

We will still run into problems however. If we use tensile steel with a strength of a hundred tons per square inch, then eventually the pile will weigh so much that we will no longer be able to pull it up, for its weight will exceed its tensile strength. The limit is around fifty miles, probably less as it will weaken at high temperature. This is not a problem however, for we do not want to pull our drill back up. The next problem is that it will eventually start to fall under its own weight, disappearing down the hole like sucked spaghetti. We will not be able to stop it without breaking it, so we will need an assembly plant to screw on more sections as it goes down at ever increasing velocity. If this became too much of a problem, we could use titanium for our pile; this is lighter, with a density of 4.5 compared to the mantle it is moving through at about 3, so we would get a lot further. Its melting point is a hundred degrees higher also. Or we could incorporate slugs of a lighter substance in the pile to achieve neutral buoyancy if we wished. Aluminium is cheap and easy to handle, beryllium is lighter still with a much higher melting point, or we could use silicon. There are dozens of lighter elements to choose from. We would incorporate it as closed cell slugs to prevent it being squirted back up the pipe, for it would not exert as much hydrostatic pressure, to borrow a term. After about 1800 miles we leave the mantle and enter the outer core. Once into molten iron however the problem is solved, for the drill will have zero effective weight, other than the tungsten section pulling it down. Eventually it will melt, if it has not done so already, and our probe will be free falling. We might as well arrange this to happen sooner rather than later; a parting membrane of lower melting point would do the trick, for once we are into the liquid rock we are well on our way. A series of probes might be a good idea, inserted into the descending pile at intervals, for if we stop it will melt, and the broken end of the pile may disappear out of sight and the hole close up due to the plasticity of the rock.

The question arises as to how fast it will fall. This depends on its overall relative density compared to the liquid it is falling through, its shape, which would naturally be quite streamlined if it has to fit down the bore, and the viscosity of the liquid, which might be quite high for the first few hundred miles. Generally, the longer the better, as longer torpedoes would go faster. The viscosity of the molten material is probably not much different from that of water, so it could go quite fast. Dense objects falling through a liquid soon reach their end velocity, when the amount of gravitational energy from the fall equals the resistance by way of drag. Probably speeds of twenty miles per hour would be feasible at a guess, but we would have to model this to get a realistic estimate.

Next we have to have a probe that can stay solid for as long as possible, and insulation and total loss cooling will be important here. Tungsten is an obvious choice, while total loss coolant could be contained in an outer jacket capable of developing an over-pressure so it can be squirted out astern like a rocket to speed progress. It could also form the power source for our high temperature electrics. Also, it will have to be able to send a signal back. Luckily sound travels well in liquid iron and rock. We will have to experiment to get the right frequency, the higher the better for data flow. Things work our way here, for most of the noise from earthquakes, crustal movement and rollers on a beach etc. are likely to be of relatively low frequency.

Instrumentation is a job for the boffins; we will need to sense pressure, temperature, magnetic field and speed, although this latter could be triangulated from the surface or by Doppler shift, which could also give us convection speeds. We also need to sample for atomic species, and probably listen to Earth noises and send these back up at a more acceptable frequency for analysis. I expect others will have a host of their own pet projects such as neutrino flux and other particles as well.

Better probes could carry more, go faster and last longer before burning out, so there is a developmental program to be undertaken. A few hundred miles might be achievable by these means.

To cut our teeth it might be a good idea to start off drilling into a magma chamber; there is a big one under Yellowstone, but practically any volcano would do. The advantage is less drilling, for the crust is thin here; under mountains it can be thirty miles thick, for mountains, like icebergs, have most of their mass under the surface. Under the sea it may be as little as three miles. A volcano gives us easy access to liquid rock to test probes in, and such sites are usually well instrumented with geophones, landlines and radio. Also much of the infrastructure is there, including roads, and cabins.

The centre is still far out of reach however, and before we get there, there is a solid core to be encountered. We could possible melt our way in given a power source. Oxygen would do the trick and the iron oxide would float free. We could examine the possibility of a thermal lance for tough rock as well; although this would not oxidise, an aluminium pipe fed with oxygen will melt anything once ignited. They are used to un-plug blast furnaces, and occasionally by safe breakers. We might build a machine to drill samples and float them back up, but all the way to the centre still looks a very tall order with current or future technology.

We will have to resort to wild cards; possible future technologies not yet invented, to explore further. One possibility would be a virtual machine. We are used to holograms, and it is possible to make a hologram with sound waves instead of light.

Imagine a hologram of a thermometer: we can see the thermometer, and read the temperature. Unfortunately the temperature will be that of the real thermometer and not that of the hologram. However, to give a simple example, supposing we projected an image of a thermometer onto a crystalline surface, where the crystal spacing was aligned with the frequency of the light falling on to it. Expansion of the surface would alter the relationship, which could be shown by an interference pattern, and we could then gauge the temperature where the image was projected rather than its source. We already have 'light tweezers' capable of handling individual atoms, so we have a proof of concept. Theoretically, we could devise a series of sonic sensors where no real machine could go. Seismography come of age! The development program might be long however. We would need to produce more ordered sound than the usual explosive charge as used in seismography. Organ pipes produce pure tones; while we normally expect an organ to be blown with air, liquids produce the same kind of turbulence at a higher frequency. We could try a tough steel organ pipe and blow it with a high-pressure water jet. Early experiments could be done with a fire pump in your swimming pool. Keep your hands out of the water though; ultrasonic sound can be highly disruptive of cell membranes. At the right frequency, the concrete lining might crumble to dust, in which case we have another method of drilling. Water jets are commonly used to remove overburden; a high-pressure water jet carrying a supersonic tone might erode rock. You could put a tungsten carbide tip on the end of the pipe and drive it round with a little water turbine: a sonic hammer drill. While we are playing with drills it might be in order to see if we can replace explosives, for we might have to excavate a chamber a few miles underground for our sound producing equipment, and again for highly sensitive geophones. Explosives are not the only way of cracking rock. We could drill a half inch hole rather than the two inch one used to accommodate a stick of dynamite, then fill it with water, or more likely a gel or soft clay so it will stay in at any angle. We then put in a washer, a soft aluminium slug that will deform to fit he bore, and hardened steel rod, and then hit it very hard. A compressed air hammer would do the trick, rather like a pneumatic drill but more so. A cylinder of a foot or so diameter and a travel of a yard, using compressed air at two thousand pounds per square inch, could deliver a couple of hundred foot tons of energy. The shock loading in the inside of the rock bore would exceed several thousand tons per square inch, which would fracture anything. Also we don't have to worry about poisonous gas, heat, dust and flying fragments as with explosives, to say nothing of handling detonators in hot rock. We might have to give our workers space suits however, for at a few miles down it would be hot enough to boil water.

Thermal insulation could be looked at. We are all familiar with superconductors; how about a superinsulator? For example, if we heat a length of copper wire in the middle, heat will spread evenly in both directions. If we pass an electric current down the wire, the current flow will tend to sweep the heat along with it. The reason is that a lot of heat conduction is due to thermal electrons. If we could find some way of locking electrons into place, or pushing hot ones out of the way so that they did not transfer heat, we might improve insulation. This technology would be of great use in industry, so we are more likely to get funding. A strong magnetic field might help. How this could be done however is anybody's guess.

Another problem might be the question of cost, and is it worth it? There are two major advantages that follow if we can drill deep holes. First, if we can drill into a subducted plate, we could deposit radioactive waste, which would take millions of years to re-surface. As long as we did not deposit so much as to form a hot spot likely to give rise to a volcano, that is. Secondly, thermal power would look a lot more attractive if we could inject water into liquid or plastic rock. We would get much higher temperatures and no requirement for fracking. Lots of minerals are soluble in superheated water, so we should keep an eye on what is coming up the bore along with the water. In the right place we might find valuable minerals, already concentrated by solution and easy to pump and separate. Go deep enough and you could have a liquid iron well. A few thousand tons per hour without the need for coke, blast furnaces, or ore separation where three-quarters of the tonnage are thrown away might be very attractive to a steel producer. With a bit of luck he might find a useful tungsten, nickel or chromium fraction as well. If we can refine our virtual instruments we could identify some very valuable sites.

However we would not have a gusher; oil wells can gush because the oil weighs about a third of the weight of the rock overburden, and it may be squeezed out. Iron has about three times the density of the rock and will not rise far up the bore. Getting a pump down might present a few engineering problems, but we could blow a lighter substance down to float the iron up. Inert gases would do the trick; the atmosphere is nearly one per cent argon so there is plenty of it. We might get iron foam at the top as the liquid argon explodes into gas, particularly if we arranged the temperature and thus the viscosity to be just right; we might find something to add to act as a foaming agent. Imagine steel beams such as are used in skyscrapers and bridges, weighing a quarter as much and twice as stiff. Handy for boats too! Again funding might be available for this, particularly if we do some experiments on foam aluminium and other materials along the way.

You may have wondered why there is so much iron. The reason is that iron is the most stable atomic nucleus. When stars burn, they begin by converting hydrogen to helium, then if they are massive enough they start to combine helium into carbon, then carbon to neon, and so on down to iron. Heavier elements than iron absorb energy when they are created, principally at the death of a star. Think of a double helix coiled spring, such as may be found in old sofas or beds. If you extend one as far as possible, like a chest expander, it has an energy store; that is hydrogen. You can let it down by stages, releasing energy all the way element by element until it is an unstressed state. That is iron. You can still compress it further, but have to put energy back in, building atoms until you arrive at uranium, the last of the natural series. Theoretically anything lighter than iron can yield atomic energy by fusion; anything heavier by fission. There is a lot of energy available when a star collapses. If small, it may settle down to a white dwarf, and nothing much more happens, as the electron pressure resists further compression just as in the earth; there is a kind of floor holding things up. Above a certain size however electron pressure is overcome and the floor collapses; this takes only about a second so it is incredibly violent. The star may now outshine a galaxy of several million stars. Then the collapse hits the basement, in the form of neutron pressure, which is a lot more resistant than the flimsy electron shells around an atom, relative to nuclear forces that is. It also has a much smaller radius. The infalling material then bounces and a shockwave propagates outwards. Eventually it penetrates far enough out to blow off the shell of the star in a supernova explosion, throwing iron and other heavy elements into space. The shockwave also goes inwards of course, and if the star is big enough the basement or neutron pressure collapses too, leaving a black hole. The early stars were big, fast and violent. The carbon they formed became embraced in later stars, where it acts a catalyst for nuclear fusion, allowing quiet stars like our sun to exist gently for a much longer time. Also, the nuclear ash produced formed planets such as ours with a mix of all possible elements, but mostly iron in our case.

The centre of the earth is not only iron though. Other heavy elements such as gold, platinum, tungsten, osmium, tantalum and a host more are there, for they would have partially settled out over the millennia, despite thermal agitation. If you would like a few billion tons of gold, it is right beneath your feet; getting it out might present a problem however. You could try selling it cheap in situ though, with free storage and security. After all, we have spent billions of man-hours digging gold up, only to put it back underground again with some very expensive security. It is only a symbol of wealth. There is an urban myth amongst bankers that some gold bars have tungsten ingots embedded in them. Nobody is going to find out if theirs is one of them however; it is the symbol that counts. More sensibly we could declare it a commons, where everybody has an equal share, say a thousand dollars' worth initially. This is only a few trillion worldwide, and we have spent much more than this on armaments. This would enable developing nations to spend on infrastructure, health and education, while the rest of the world was provided with lots of jobs. While we are at it we might as well declare the atmosphere and oceans a commons too, where polluter and exploiter pay into a common pool.

Some might question whether there was any gold was there; supposing it was all tungsten? Well, if it does not matter on the surface it certainly will not matter down there. It is the symbol that counts. Since there would be so many gold certificates about we could use the real gold to make pretty coins, which would be gradually lost over the centuries, going back into the ground again, a sort of natural cycle.

Having had a look at some wild card technology, and there could be a lot more, we can go a bit further and look at science fiction. After all a lot of this, though impossible at the time, is now common place.

I will leave it to you to trawl your favourite stories for leads.

COMMUNICATING IN THE

MULTIVERSE

We now all live in a multiverse, or so we are told: an infinite number of them, or maybe just 10 to the 500th, which seems pretty close to me. This is really no more than an extension of a trend that has been with us for some time. First the earth was the centre of the universe, then the sun, which later gave way to being just one star in the galaxy, then our galaxy paled into insignificance being just one of many. We have exchanged the conceit of the earth being the centre of the solar system, for the greater conceit of being at the centre of everything. Admittedly, in this age of equality everything else is at the centre too, a kind of variant on a Caucus race, a la Alice in Wonderland.

Personally, I prefer to skip a frame and have a multiverse of multiverses; they come cheaper by the 10 to the 500th to the 500th, if that is the packet size. This is still not infinity though; perhaps we could have it going on for an infinite time; that should do the trick. Infinity is, after all, a direction, not a destination. There must be some giant lookalike M&M or Smarties factory the other side of reality, churning them out like billy-oh. For me, it is either infinity or nothing; there is no point in stopping half way!

They say we cannot possibly communicate with other universes in the multiverse. I disagree. It could be done by using implicate quantum pairs, for they are not pairs, but one single item until they are split, therefore there is no time or distance between them until that point. You could even get information from inside a black hole, if there is any there, that is; it might all be stuck on the event horizon. The trick is to have a stack of quantum pairs split between the receiver and the sender; by forcing the collapse of one, one can set up a binary code. We might have a problem sending the other half of our quantum pairs to another multiverse, but luckily we do not have to. Somewhere there is bound to be another world just like ours, with someone else setting up identical pairs, in which case the job is done for us, for if they are the same, there is no difference, so to speak, and our pairs will be implicate with theirs. If you don't do quantum, think morphic resonance, or possibly synchronicity. After explaining all these carefully to a friend, he came up with a further possibility called 'Du Wha?' to which I agreed, not being familiar with the subject. He also believed in the transmigration of souls, although in his case, I suspect emigration.

I try to tell people that there is another parallel world, with parallel people out there, which we can communicate with. It explains why the psychologists have not come up with any explanation for the mind; they are looking for it in the head, whereas most of it lies out there. This apparent sense of self is really the constant chatter between multiverses of yourself, with the occasional random fluctuation that leads to new ideas. (I wish I had thought of that!) In fact there is an infinitude of lookalikes, being a subset of the infinitude of multiverses, for any fraction of infinity is also infinity.

There is not one copy, but a multitude, all shifting in and out of focus as random effects shimmer the veil, like a picture projected onto a cloud. You are not much different yourself; there is not much left of the 'you' of seven years ago, apart from tooth fillings, but you are still you. People tend to ignore my reasoning though, despite it being universally agreed by me. I might just as well be talking to myself.

You can of course alter your universe For example you could move the footstool a foot to one side; although trivial, this is a different universe. Somebody else might trip over it. Best not to explain about different universes, though; they might wish you into another one, particularly nasty, and a long way away. I am not so sure about the distance though; universes are not separated by normal distance, but a kind of alternative dimension. Maybe though they only exist in the mind. I am quite happy with that however; after all, that is the only place I live too.

Just occasionally though, I do meet someone who knows. Only the other day I was sitting opposite an exceptionally pretty girl in the cafeteria, a high flyer in cryogenics, I was later told.

"Do you know," I began, "that in a parallel universe I met a girl just like you and asked her out, and she said yes?"

"Ooh!" she said. "How exciting! Another universe, and we are together in it! How thrilling! So you won't mind, even the teeniest bit, if I say no in this one?"

Smartass.

Come to think of it, the other girl had a much nicer personality.

SECTION 7

SCHRODINGER'S CAT REVISITED

Here we re-visit some old puzzles, with a few new ones thrown in for good measure, by way of mental exercise. Good for burning off calories too! The brain uses about ten-per cent of the energy of the body, so you don't have to go for ten mile jog if you don't want to.

In this modification of Schrodinger's cat in a box experiment the cat is in the box, with a radioactive source and a Geiger counter as before. If the radioactive source emits a particle that is detected, a circuit is completed that releases a poison gas, killing the cat. For the experiment we will use a source that will, on average, complete the circuit every ten hours, and the duration of the experiment will be five hours, giving a 50:50 chance of the cat being alive or dead at the end of the experiment. The box is in a sealed environment where no information can get in or out to affect the results.

The modification consists of a stop clock in the circuit, set running at the correct time at the beginning of the experiment. This, to a degree, makes the cat and gas redundant, so cat lovers may leave those out if they wish, inserting a lamp that can be off or on for dead or alive.

As we have no information until we open the box, we can say that the cat is in a superpositional state, both alive and dead until a measurement, that of opening the box, collapses the probability state and it is either one or the other.

There are two questions to be asked:

1) How superpositional is the state? If we open the box after one second, it is highly unlikely that the cat will be dead. The probability is one in 36,000. This decreases to 50:50 after five hours. We could argue that the state is always 50:50, but this presents a problem if we forget to open it five hours later, or open it before, for we then have a future event affecting the present, a kind of reverse time causality.

This idea of a fractional amount of superpositionality, which changes over time, is a new element in the experiment.

2) When we open the box, we collapse the probability state. Thus, if we had set the experiment running at 10 in the morning, and opened the box at 3 in the afternoon, the state would have collapsed at that time. However, in a real experiment we find that the cat is dead, and the clock has stopped at 10-30. So when did the probability state collapse? When the box was opened, or at 10-30? Or did it collapse twice, once when we opened the box and saw a dead cat, then again when we saw the clock had stopped at 10-30? The idea of a probability state collapsing twice is new. We could argue that it was always 10-30, but what if we forget to look at the clock? Perhaps we should make a distinction between a state, and knowledge of a state, the knowledge being a secondary derivative of a pre-existing state.

There are other variations we can add to the experiment, such as branching causal loops, which feed back to some point in the chain, but I will leave that for others to explore.

PROBLEMS AND PUZZLES

Many people like problem solving. It is good mental exercise and can give a sense of achievement. Even failing to solve a problem can be a learning curve.

There are several categories of problems. For example there are known insoluble problems, like the trisection of a given angle by straightedge and compass only, or the three body problem, dealing with the orbital paths of a three star system. Sometimes problems believed to be insoluble are eventually solved. For example, the trisection problem can be solved by cutting out the given angle, rolling into a cone, inscribing the circle formed at the open end onto a flat sheet, and then trisecting the circle by stepping out the radius; this can then be transferred to the cone and unrolled.

Cheating should always be considered in the face of a Gordian knot.

The most important kind of problem is where there is no known solution, but probably is, if someone can work it out; work in progress, so to speak. The solution of Fermat's last theorem is an example of a recently solved one, although Fermat never solved it that way. Sometimes there is a hidden solution, but not widely known, so publishing the problem may bring in a solution from an unexpected direction. The realisation that the maths relating to entropy is the same as that for information is one example, with the result that instead of a universe that is 'decaying to entropy' we can just as easily say it is decaying to information. Complexity increases, just as order increases with cooling, according to the third law of thermodynamics.

I include a few problems, some with solutions, some without, to get the page rolling. No doubt others will present better ones later.

1) Given that the early Egyptians divided the circle into 360

degrees, using a straightedge and compass only, how did they do it?

They could construct an angle of 60 degrees, by inscribing a six-sided figure in the circle, and this could be bisected to 30 and again to15 degrees. They could also inscribe a pentagram, giving 72 degrees, which by bisection gives 36, 18 and 9. By overlapping with 15 degrees they could then get 6 and 3 degrees, but it is impossible to trisect an angle. It could be done by inscribing a nine-sided figure in a circle, to give 40 degrees; bisect down to 5 and then overlap with six, but I don't know how to inscribe a nine-sided figure. You can try to find a solution.

2) Next, you are asked to perform an experiment that consists of firing small pellets from a little air-gun, at an object in the dark which is suspended from a wire, which is illuminated so you can see where its centre of gravity should lie. The pellets emit a flash of light when they hit, which is automatically recorded on a monitor screen, to build up an accumulated picture. As this will take a little time, you are paired with another volunteer, and you decide to take turns. After an hour he presents his results:

'It seems to be an oblong, about four by eight inches,' he says, 'but half of the pellets seem to go through, and are reflected by another, eight inches back. Also, the rebounding pellets leave little streaks of light for a few centimetres. From these reflections I assume it to be cylindrical, possibly like a squirrel cage from a motor; that is, a cylinder perforated with slits.'

You take your turn, but being idle, unscrew the barrel of a Bunsen burner that is handy, slip it over the end of the air gun, tip in half an ounce of shot and fire the lot at once. Surprisingly, you get a four-inch circle instead of an oblong. Thinking maybe the shot is constrained by the added barrel and you have a shot pattern, you slip it further down so that there is no choke effect, and try again. This time you get a vertical bar to the right of the picture, four inches by half an inch. Another try gives you a vertical bar to the left. You are not asked to explain your results.

3) Next, you are shown a glassed cage in the middle of what looks like a skating rink. You are told that it is ice, but coated with a very slippery polymer, so that it has an extremely low coefficient of friction. To demonstrate, an assistant puts a clockwork toy car on it; the wheels spin, but it fails to accelerate. Your problem is to return to

the side having been placed in the middle of the pond. You may ask for any reasonable piece of equipment, so long as it does not exert an influence outside of the cage. Rockets and large magnets etc. are out. This is known to be impossible, being a 'non-Newtonian' device; that is, one which can proceed outside of the space previously occupied by it with no Newtonian reaction.

There is one thing you might try though. The object in the shot experiment was a precessing gyroscope. You can try experiments on precession yourself. If you don't have a gyro, you can make one with a bike wheel; take the tyre off and put a bit of lead pipe, well secured, in its place if you want to beef it up. It helps to weld a length of builder's steel rod to the spindle so you can hold it better. Bike wheels make good pulleys too.

If you spin up the gyro, and hang it on a cord so that it is at right angles to the cord, it will precess. It will go round in a circle, with the direction of travel of the top of the gyro in opposition to the movement round the circle. The cord hangs vertical, as the gyro exerts a torque, bringing the effective centre of gravity under the point of suspension. This is its response to gravity; increase the apparent gravity by accelerating it upwards and it will speed up its precession, lower it and it will decrease.

If you now drop the gyro, precession will stop as soon as the cord is released. As it strikes the floor you can mark where it struck (use a cushion, a low speed of rotation and someone to catch to be on the safe side!) If you then repeat the experiment, dropping when the gyro has precessed 180 degrees from the previous position, you can make another mark.

Interestingly, although the centre of gravity is always below the suspension point, the centre of mass is not. You can choose where you drop the mass on the circumference of the circle described by the precessional path.

You could try asking for a gyro in your cage on ice. Precess the gyro so that the centre of mass is towards the side of the rink you wish to approach, and drop it. Then move the gyro back to the start position and repeat. Now the movement of the gyro under precession does not give a Newtonian reaction, as is shown by the string hanging vertical while it precesses. However, moving it back not precessing does; it is a mass as any other, so there should be a reaction force, causing a movement, in inverse ratio to the mass of the gyro and of you in the cage as you reposition it. It might be easier to try the experiment with 'three men in a boat,' in a swimming pool, where the gyro is passed precessing from hand to hand one way down the boat, and back, two handed and not precessing, the other way. I do not know if this would work though.

If it does, you have invented a useful 'space wheelbarrow', which would avoid the use of rocket backpacks in proximity to optically clean surfaces on a space assembly site. Don't try to cross the universe with it though. It might take a long time, as the thing won't accelerate!

BREAKING A ONE-TIME PAD

Some problems are known to be insoluble, others presumed so. However I believe there is merit in attempting to solve impossible problems, not so much in hope of a solution, although that sometimes does happen, but as an intellectual exercise, and aid to thinking outside the box. Also of course you never know what is likely to turn up in the by-lines!

Cryptanalysis, sometimes called code breaking, is an intellectual pursuit of a high order. Many people do it without knowing when they attempt to solve a crossword puzzle. Translating from a foreign language is another example.

Professional code-breakers, or more accurately cryptanalysts or cipher breakers, are more concerned with trying to read secret information encrypted by another state, with a view to gaining an advantage by way of finding out their intentions prior to action.

The words code and cipher are often used interchangeably, although strictly speaking a code is a substitution of a word or words for another, while a cipher is the jumbling and or substitution of the individual letters of a text.

The gold standard in the cipher world is the 'one time pad', or random number string. It is completely unbreakable, and this was proved to be so in 1945. I have not read the proof, as I tend to be idle and agree too readily with other people's findings.

In its simplest form the message to be encrypted, called the plaintext, and is converted to a string of numbers, with A equalling 1. B is written as 2, and so on to Z, equalling 26.

Below this number string is written a completely random number string. It is important that this string is truly random, not some algorithm; the extension of Pi or the golden mean will not do. They are too obvious and not truly random, although of course you might generate the first few numbers of Pi by accident if you pick numbers out of a bag. A completely non-random sequence can occur randomly; will do in fact, if you keep at it long enough. It can be difficult to decide if a string is truly random, but there are a few tests, such as the distribution of doubles or triplets. If you try to invent a string you are likely to avoid these. Also, if you bash away at your keyboard at random you are likely to favour some letters above others. I generate my random strings by taking the twenty six letters from a Scrabble set, shaking them up well in a pot, then putting my hand in and scrunching them up a bit more before taking one out, writing it down, putting it back in and repeating the process. It will have to do! Ideally one should have a grid array of twenty-six letters, with a sensor in each square to spot cosmic rays, and an automatic recording device.

The plaintext number string and the random one are then added together, resulting in what is called the cyphertext. If you do not have the key, which is the random number string, you cannot break it, for the message plus the random component is also random.

The only way to try is to perm all possible combinations of random number string and subtract from the cyphertext. Eventually you will hit the correct one and generate the original message. This may take rather a long time, but you will be certain to succeed eventually.

Unfortunately along the way you will also generate all possible messages that could have been enciphered, including all the Shakespeare sonnets plus those he never got round to writing. You end up with too much information. It is a bit like having a book where all the pages have been printed onto one sheet of paper. It is all there, but the page is black!

You must never use the same number string again however, for if you do, somebody may subtract the two messages from each other. This means that the random number string has disappeared, and all you have left is one message added to another. This is breakable, with difficulty, for if you again perm all combinations the two messages will drop out in clear amongst the clutter of garbage and single messages. This is of course why it is called a one-time pad.

So if this is the one and only truly unbreakable cipher, why is it not used anymore?

The problem is that it is cumbersome; not so much the adding up, but because the weak link is transferring the random string to the recipient. It may be intercepted, or copied in his office. This is not much of a problem apart from the time taken to fly it maybe halfway round the world, but the problem gets bigger if there are thousands of people involved. There is a logistical problem in moving stuff around, and with every copy the risk of a security break increases.

Also, because it can only be used once, everybody in the loop has to delete the used section even if they did not use it. If it is one embassy talking to another there is not much of a problem. If you have several thousand tank commanders talking to each other and area headquarters the problem becomes exceedingly difficult.

Nowadays the preferred method is to use big number codes. The big numbers are huge primes. It is very quick and simple to multiply two huge primes together, but a problem of an entirely different order to factor them. It is a sort of one way gate or function. It can be done of course, but it takes an awful lot of computer time. Computers get faster and faster, but one can find bigger primes even quicker. Of the making of prime numbers there is no end; if you want a bigger prime than the biggest one you have got, all you have to do is to double it, add two, and the resultant number will be the sum of two primes. One of course will be bigger than the one you started with. If not, then you have broken Goldbach's Conjecture (any even number is the sum of two primes), and you will be famous. It is a conjecture only, as it has never been proved, but not disproved. Similarly, any odd number can be formed from the sum of three primes, the simplest case being Goldbach's two primes, plus the number one.

You still have to find your bigger prime of course. It might be easier to try for the smaller of the two first. When you have them, don't stop, for there will likely be other solutions with different pairs.

If there is no end to primes, there should be no problem looming for big number codes, but there is: two in fact. First of all, big numbers eventually get unwieldy. If I said I was going to e-mail you a prime so big that if written out at ten characters to the inch, it would stretch all the way to the nearest star, you would probably say 'thanks but no thanks!'

Even worse there is a new type of computer looming, which uses quantum bits or 'Qbits' as elements of calculation. In ordinary computing a gate is either open or shut, giving either a 0 or a 1. A 'Qbit' is a yes, a no, and a maybe all at the same time. Also it is implicate, or mixed up with one or more other 'Qbits', and the collapse of the state of one triggers the collapse of the state of the others or some of them in strings, according to how you mixed them.

The net result is tremendous parallel processing power. If you can imagine a few square miles of graph paper covered with numbers, and yourself in a helicopter with a telescope trying to find a couple of primes among this sea of information, you have the problem of an ordinary computer. If you could shine an ultra-violet light down on your number sea, and two of them glowed brightly; you have the problem solving ability of a quantum computer. Big primes will become insecure.

The obvious answer will be to use quantum computers for communication. There are some problems here however. To start with, they don't exist yet, or not in marketable form. Then nobody knows how to program them, or at least there is no body of knowledge or existing protocols to do it. Also, the hackers will have them as soon as you do, and another arms race will develop.

The problem is one of a gap, a stretch of time where current technology is known to be insecure, but nothing is up and running to supplant it. One big advantage of quantum computing is that it is easy to spot if someone is trying to hack into a message. Quantum communication will be totally secure! Or will be until someone comes up with a way to break it. There are already hints that it may be possible to intercept a quantum message unseen. The problem is that quantum computers can still work when switched off. The quantum world is weird, and it may be possible to see what doesn't happen if you don't do something, rather than doing something and seeing what happens, by using a virtual photon. It is a bit like that problem where you are in a room with two guarded exits, one leading to freedom and the other, death. One guard always lies and the other always tells the truth. You are allowed to ask one guard one question to find the way out. The secret is to set up a double negative question. I won't spoil it for you, in case you have not met it before, by telling you the answer.

If a random number string can be transmitted securely, then we could go back to using this one unbreakable system again. Also we could have any number of dedicated random strings for different operators if we wanted, thus obviating the distribution problem and the security problem where many people are using the same one.

Thus there might actually be a case for examining the random number string to see if it really is totally secure. At first glance the problem does seem to be impossible; this means that you are unlikely to have a peer group. Even if you are lucky enough to have two others interested, one will probably have doubts about your sanity, and you may have even greater doubts about one of them. This is only the median; if you perm the possibilities, you will see that it could be worse. On the up side you will not have much competition, and you will always have work in progress, something to pick up and put down again as to mood takes you. Also, you cannot really fail, unless you succeed, that is, for to work on an impossible problem and not solve it is can hardly be called failure.

Code breaking is a matter of attrition. A lead will reduce the number of possibilities to be explored, a bit will be pried off here, a clue will arise elsewhere, and the problem starts to reduce. It is rather like quarrying a mountain. You don't do it all at once!

We can thus look at different elements of the problem. Alan Turing proposed a 'universal calculating machine'. This was not a real object, although it led to the programmable computer which we love to hate today, but a concept.

In a lighter vein, I would like to propose a, 'universal failing machine'. This is not a real machine, but a concept which can take any form, its main function being to go wrong, and thus explore what can go wrong with a system.

We will propose a concrete example to deal with a real code. In this machine the input, the plaintext, is typed in on an ordinary daisywheel typewriter, if you can find one. It is modified to also print the number equivalent of the letter underneath.

Next to it is another similar machine which prints the random code from a tape, again with the number of each letter printed underneath so you can see what is going on. The two numbers are added by a simple electronic circuit on a chip from an old computer, and sent to a third machine, which prints out the cyphertext.

I now type in my secret message to be encoded: 'The quick brown fox jumps over the lazy dog.'

The print comes up with numbers underneath. The random tape prints in synch with numbers underneath, and the cyphertext prints up, 'The quick brown fox…'. Something has gone wrong! I look at the cyphertex printer and identify the problem. The daisywheel has two drives; one prints the numbers of the plaintext, and the other drives on further to add the random string. If the total is more than twenty-six it merely runs on and starts again (called modulo twenty-six, which I will come back to later).

However, the random string drive is slipping on the spindle; there is a grub screw with a crack in the plastic and I was wary of over tightening. I cautiously tighten it up a bit and try again. I now get, 'Thf qujk crown fpx …' I can see what is wrong; it is still slipping, but occasionally moving one space and adding a one to some of the letters.

This is hardly well encrypted. We could say that the signal to noise ratio (of which more later) is far too good. So I tighten it a bit more and find it is adding some ones and a few twos. This is a bit better encrypted, but any codebreaker would laugh at it. So I tighten progressively, getting a bit more encryption each time. Finally I tip some superglue on it and solve the slippage problem.

However this has exposed a different sort of problem. The message gets harder to decrypt each time I tighten my grub screw, but where is the tipping point where it goes from being merely difficult or very difficult to becoming impossible?

There is a fundamental difference between very difficult and impossible. The difference is qualitative, rather like trying to reach infinity by counting one, two three, etc. To put it simply, you can't get there from here! I could say that with only one step working it is one twenty-sixth impossible, then two twenty sixths until I finally reach twenty-six twenty-sixths impossible. That however is about as good as saying one can have a twenty-sixth of infinity.

However it is not actually as good as twenty six twenty sixths. Every time a Z comes up in my random number string I add 26 to get back to the letter I started with. In other words, very nearly 4% of my cyphertext is actually plaintext. So I only have twenty five twenty sixths of impossible. This is not actually a great problem; a good code should allow a letter to encode as itself. One fault with the Enigma machine is that it could never do that, so at least you knew what a letter wasn't! To point this up, let us suppose that we have a machine that cannot encrypt a letter as itself, and we then transmit a binary code. This only has 0s and 1s, and every 0 becomes a 1 and every 1 becomes a 0. Hardly a very high level of encryption!

However, the fact remains that on average one letter in twenty-six is the same as the original, even if you don't know which ones. Back tracking, we could then ask how this is qualitatively different from a system where two letters in twenty-six are not encrypted, or three, or four. You might say that this is a totally different problem, and of course that would be less secure. However one in eight letters are likely to be 'e', as this is the frequency with which it turns up in most prose, so a few of the 'e's are likely to correspond to the plaintext as well.

We will leave this for the moment, but we have uncovered a slight logical contretemps. This should ring a faint alarm bell. It may be a false alarm, but it is worth making a note of it.

MODULO TWENTY-SIX

If you say to someone, "I will meet you under the clock at Waterloo Station at two o'clock," they will know what you mean, and are unlikely to be there at two in the morning. If not, they will hang about in the roadway with little luggage trains dodging them for twelve hours. If someone is in the services they will probably say 'fourteen hundred hours', using a twenty four-hour clock. They don't actually mean fourteen hundred hours; that would be nearly two months later. Also, the two zeros actually stand for minutes, which are sixtieths of an hour and not hundredths, but people will know what you mean. Sometimes you will see a twenty-four hour dial in Europe. It is rather congested however, and would not look well on a small wristwatch, so we use twelve hours and then roll round again. In adding our code numbers, there is no twenty-seventh letter of the alphabet, so we go round again. We call this modulo 26. Numbers in modulo can present problems if you don't know what modulo is in use, particularly if you divide or multiply. However there is no real problem in writing down your numbers, as we know we are in modulo 26. You can leave them unconverted and sort them out later if you want. Modulo 26 works just as well backwards and the clock unwinds when you subtract. You might ask. 'Why not say Z is worth zero instead of twenty six?' I have left it as twenty-six for clarity in some graphs that follow.

So in cryptanalysis we know the numbers, and the letters derived from them are in modulo twenty-six, right?

Wrong, actually. Half of them are not. If your random number string throws up an A for 1, then all of the plaintext letters it can be added to will be less than twenty-six, except Z, which stays the same anyway. If you have a two in your random string then all the numbers below 25 will not get as far as being in modulo 26.

Even this is not quite true. More than half are likely not to be in modulo, for if you take a reasonably long stretch of plaintext, add up all the numbers and then divide by the number of letters, you are likely to find the average is a bit below the average for your random code. Some letters are used more than others; ABC are more common than XYZ for instance, and e is the most common letter of all. About one in eight will be e instead of one in twenty six as for the random string. The result can sometimes be a long way out. As an example convert the instruction 'add all', to numbers and then take the average. It is only a bit over five instead of thirteen and a half for the random string.

Again, we will leave this, but we have exposed another interesting fact about modulo numbers.

TRAFFIC ANALYSIS

Even if you cannot decipher a message, the number and frequency of messages can tell you something. If for example at two-o clock Sunday morning the main cyphertext facility in Utopia suddenly lights up and starts firing off messages all over, you may guess that something is up! But what? Have they had a reactor meltdown? Is it a coup? Are they starting to invade? So you try to wake up agents on the ground to get some clue as to what is going on, which you would not have done if you had not monitored for unusual traffic.

If we can have unlimited random code, we could beat this telltale by having continuous transmission. The code would run on letter after letter, the decoder would subtract its copy and give an unending run of zeros, until a message was added, that is. Then the message, stripped of its random element would stand out in clear. The traffic analysis boys would be none the wiser; all they would get would be a stream of gibberish whether it had a message encrypted or not.

Or would they? If you add a string of numbers with a lower average to a random string, the end result is likely to be a bit lower than expected if you divide by two. Modulo twenty-six might mask it a bit, but that is incomplete, with less than half getting into modulo.

To point this up, take a random string of numbers, add them up and divide by the number of numbers and you should get near to thirteen and a half. Now take the same number string, and add a series of ones to it. This is an extreme case, designed to show up any variation; for a real text message the trace would be less obvious. If you now take the average and halve it, you will get nearer to seven. In other words you could tell if you had just a random string or one with a message embraced.

So traffic analysis might still work by reason of the difference between a truly random string and a text, which favours some letters over others. In other words, a meaningful string plus a random string is not necessarily random.

NOISE TO SIGNAL RATIO

All signals contain some noise. This is why digital transmission is preferred; it cleans up the signal, and even if one or two digits are occasionally wrong the rest are still clear. Eventually even digital will break down, when the noise is so great that you get random digits popping up all over the place as the digitiser tries to make sense of the fuzz.

The noise to signal ratio for a one-time pad is near enough one to one. This does not seem a very high ratio, for you can hold a conversation in a noisy street where the ambient noise is louder than speech. Actually it is a bit below one to one, for text has a front end loading for smaller numbers, and some of the ciphertext is actually plaintext in disguise.

Strangely, noise can often enhance a signal. It is time for another universal failing machine to show how this can happen. Let us suppose we are trying to receive a signal from a distant space probe. This one is a solar sail, several light-years away now. The problem is that every time it doubles its distance it needs four times the power to give the same signal strength at the receiver. This may be death to SETI, for if extra galactic intelligences communicate by radio, they will need big transmitters! Some distant stars only yield a photon every second or so; they have to be stored for hours before we can get an image. Yet these stars are burning a thousand tons of hydrogen a second, just to give us the odd photon or two.

Our sail is parabolic, and when transmitting, it focuses on a solar panel to provide power, but not when not transmitting, as this would reduce the sail's thrust. Every time it doubles its distance the available power is only a fourth of what is was before, so one hour's worth of transmission becomes about four minutes at double the distance. Eventually, the signal drops below the threshold for detection.

You can show this in a graph. The simple sine wave is the transmission. The horizontal line is the audibility threshold. The ragged graph line is our old friend, the random number string, here representing noise, partly from stellar hiss, partly from jostling atoms in our apparatus, even though it is cooled with liquid helium.

The other bold irregular line is the signal and the static combined. Hey presto! Bits of our signal reappear where the random noise reinforces the signal. Like some ghost from the dead the signal is revived, albeit in a rather erratic form.

We will go back to our original failing machine for another example. I type in the instructions: 'add 1.00 times plaintext to 1.00 cyphertext, and print numbers.'

I type in 'The quick...' The numbers come up; 20 8 5, 17 21 9 3 11, as these are the equivalent alphanumeric positions of the letters of the alphabet.

The random string comes up; mfw jrds, and the alphanumeric of these letters; 13 6 23, 10 18 4 19.

Then I get the ciphertext, the sum of the two strings: 2013 806 523.

The failing machine has failed again. Instead of 1.00 times plaintext, it has lost the decimal point and is adding 100 times the plaintext. The signal is a hundred times greater than the noise, which is completely swamped by it. It is hardly encrypted at all, for we only have to reinsert the decimal point, and the plaintext is all that is to the left of the point, i.e. 20.13. Interestingly, we can also recover the random string, as that is to the right of the point.

I retype the instructions and repeat the process. The numbers come up as before, but this time I get 1320 608 235. Another failure! This time it has omitted the other decimal point; now the noise is a hundred times stronger than the signal! At first take one might think that this is a very strong form of encryption. However, the two strings can be separated quite easily merely by putting the decimal point back where it should be. This is rather contraintuitive. There is obviously more to explore regarding noise to signal ratio. Perhaps for alphanumeric strings at least, 50/50 is the best ratio.

Of course, it is easy to see what is happening when one has both the plaintext and the random string to play with. However in the real world one will only have the ciphertext. Could we however alter the noise to signal ratio? We could perhaps add more noise, perhaps add another random string of our own, or take one away. Perhaps we could add 99 strings and take away 100. If we can alter the signal to noise ratio away from 50/50 we might weaken the cipher. We can even increase the signal strength by doubling our numbers; if we have reduced the noise level, this will boost the signal strength without boosting the noise proportionately.

DISTANCE OFF AND SPREAD

When navigating a coastline in a small yacht, it is important to know your distance off the shore, so as to avoid hidden dangers such as outlying rocks, reef or sandbars.

This same idea can be applied to code breaking, except that here we want to be as close as possible to the original. Your cyphertext may have any number from 1 to 26 added to it, the average being thirteen and a half. If we can reduce this average, we have reduced the random element, rather as our universal failing machine did. If we can reduce the spread of numbers down to three or four, that is, the cyphertext only being removed by that amount from the message, we are very close to cracking it. We will have reduced the noise to signal ratio, so that the signal stands out over the noise.

I will show how this may be put to use in the next section.

HOW TO BREAK A RANDOM
NUMBER CODE

In this code, the letters in the message are converted to their alphanumeric (the place position in the alphabet, with A being 1, B 2 and so on). This is then added to a random string of numbers and reconverted to text.

I have seen this called an impossible problem, the reason being that when a plaintext message is changed into its alphanumeric equivalent and added to a random number string, the result is random.

This is not quite true, for while the average alphanumeric for the alphabet is thirteen point five the average for something written in English is a bit below twelve, as E and A occur frequently in English and are low numbers.

Thus it is possible to tell if a number string has no message added and is truly random, or if it has an added message, which will depress the average value to below thirteen and a half. However, this is easily avoided by 'packing' the message with a few spurious high value numbers. A few XYZs scattered in the message will be easily recognised as packers and discounted. More realistically, don't use XYZ but a mix of high numbers individually scattered.

I have also seen it stated that it has been 'mathematically proven' to be impossible. However all proofs have premises, and some assumptions are more valid than others. All problems are impossible until solved.

The code can of course be broken by brute force. To do this, all possible permutations are worked through. Thus for the first letter of the cryptext there are twenty six possibilities, A to Z. The next letter is permed against each of these possibilities, each one generating another set of twenty-six. Numbers obviously soon become astronomical and then not calculable in a reasonable time.

For a very short message it would be possible however, but as well as generating the original message (amongst all the garbage) one would also create all possible messages that would fit within the length, with no way of telling which is the right one. The method therefore does not work.

We will next examine a couple of special cases, where there has been an encoding error that enables the code to be broken.

The first is where the same number string has been used for two or more different messages.

Here, the subtraction of one message from the other will remove the number string, and all one is left with is one message encoded against the other. If permutation is now attempted, the two messages will drop out back to back in clear, and other messages that only occur once can be discarded. This is still very long winded, so in practice it is better to try out common words, such as 'the', assuming that every three letter block is 'the', in turn. When it does coincide with a real 'the' another word or part thereof will drop out in clear. There are also a lot of other words, from 'theatre' to 'they' that will be exposed also. Next another common word, such as 'and' could be used. Before long the message will start to take shape, and before half of it is retrieved it will be possible to make guesses as to the rest.

However the chances of such a mistake being made are remote, although it has happened. If you do happen to get one, it is probably a plant; the next real one may be several years and a million messages later. So this is not serious codebreaking.

The second encoding fault is in a sense the obverse of the first. Here, instead of the random string being used more than once, the encoding machine in the Dystopian Embassy goes into a loop, and instead of shutting off at the end of a message, starts at the beginning again using a different number string. Thus we have repeat messages, all the same, but different in cryptext because they have a different random string added.

Permutation will not work here, for although we can generate messages as before, we have no way of recognising the number string it is paired with, as this is just a random string like any other, making no sense as a back-to-back message did.

If we subtract, the message disappears, leaving the difference between the two random number strings. It is instructive to see what information can be gleaned from this. For example, the number may be odd, in which case we know that of the pair, one must be odd and the other even, but not which. Similarly, if even, we know that the pair are both odd, or both even, but again, not which, this halves the possible numbers of combinations, but this is not a lot of help. Given a large number of random faulty messages, one can try other combinations. Put in Pi or tempt it with primes if you like. I will leave this to you and move on to another way of breaking the code.

This is done by reducing the noise/signal ratio. To do this we add together a number of faulty messages, the larger the number the better, but suppose we start with ten. We now have ten times the signal strength, but, unfortunately ten times the noise also.

However, noise is to a degree self-cancelling. For a long number string the average will be close to thirteen point five, as this is the number you get if you add all the number equivalents of the letters of the alphabet and divide by the number of different numbers you have used.

You can show this graphically; to do this, on the vertical axis, write down the alphabet on one side of the line, and the alphanumeric on the other. Draw a line to form the horizontal axis under the number thirteen.

Then put in a section, say ten letters of your random string. Join up to form a graph and you will find that it zigs and zags in a random manner across the page.

Next, start a second graph in a different colour; for this one start in the same place, but for point two, add point one and point two of your first graph, and divide by two. For the next point add the first three and divide by three and so on until the end of the graph.

You will find that the zigs and zags rapidly diminish, closing in on just under the horizontal axis. The further you go, the closer to thirteen point five it will become.

Therefore we subtract 135 from our total in each column made from our false messages as above. We have now reduced the noise. We then divide the remaining number by ten (because we have ten times the message) and translate to the nearest alphanumeric. We can put another line above this for the next lowest alphanumeric and one below for the next highest. Most of our original message will straddle these three lines. Permutation of three lines is a much easier problem than permuting the whole alphabet, and the message will mostly stand out anyway.

We will now do a worked example. To keep it short I will use AND, as our message, for what is true of one word is also true for many words.

The cryptographic machine will have rendered this into alphanumeric.

A N D

1 14 4.

Then add a random number string:

5 8 17

To get:

6 22 21 which is our cryptext.

13 26 6 This process is repeated another nine times.

4 29 22

18 37 20

26 22 23

20 35 14

11 27 23

17 24 8

24 22 14

4 26 14

143 263 165, then add the columns:

135 135 135, then subtract 135:

8 128 30 and the result divided by ten.

The nearest alphanumeric is:

8 12.8 3, substituted:

A M C, then insert the line over and the line under:

B N D

A M C Thus the most likely word is AND

Z M B

You will have noticed that I have put Z for the letter less than A in alphanumeric. This caused by 'Modulo 26'. When the total of two numbers exceeds 26, it rolls round the alphabet again, rather as a clock goes from twelve noon to one pm.

Also of course, you will notice that most of the numbers in column two are above 26, and, in text would have rolled round, giving you a false set of numbers. There is no point in doing it here, for what rolls round in one direction, rolls back again in the other.

There is a way of attacking modulo 26 which I did not put in at that juncture for the sake of clarity. I will now explain how its effect can be in part negated.

Only half of our letters are likely to be in modulo, or a few less if letter frequency is taken into account.

Suppose we have A in the cryptext, we ask the question, 'how likely is this to be in modulo?' Or in other words, 'what two numbers of the alphabet will add up to one?'

There are none, so the chance of it being in modulo are 26/26.

Then we examine B; this could be a combination of A+A, and no other, so the chance is 25 /26.

We then examine C; this is A+B.

We work down through the alphabet, ascribing statistical probabilities to each letter, down to Z. This can only be Z+Z or Z+Z to put it into modulo so the chances are 25 to one against. All other letter pairs that add up to 26 are below modulo.

Thus we propose alternatives to these letters at the time of addition. Actually, with a very long series of faulty messages, modulo is self-cancelling, for when we divide our final result we also divide the modulo element. Thus if we have ten letters to be added, half are likely to be in modulo; this subtracts 135 from our total, which we were going to take off anyway.

So we can break this special case where a cypher machine repeats the same message with a different random number string each time.

So what are the chances of that happening? Probably about zero. So the method, although interesting, is of no use in codebreaking.

Next we see if we can convert a special case to a general case.

Suppose, as is usual, we have only one cryptext instead of many? Can we create spurious repeats? In other words, if the Dystopian Embassy is not going to oblige by encoding the same message with a stack of different keys, can we do it for them?

We can easily generate a series of spurious messages by messing with the Dystopian code, We could for example subtract a line of code from the cyphertext. It will be the wrong one, of course, but that does not matter, for we can then add another, subtract another and so on, and generate a sufficient number of variant cyphertexts for our purposes. Suppose we generate ten variants, we can treat them in exactly the same way as we did for our worked example.

There is a problem here however. I can explain this by means of a fiction. Everybody who has hitch-hiked around the galaxy knows that the answer to the riddle of the universe is 27. The problem is that nobody knows the question. There are an infinite number of ways of arriving at the number 27.

Normally, addition is considered as a one way function. However, let us suppose that when you add two numbers together they do not merge like milk into tea, but instead retain their individuality, like currants in a bun, and your plaintext and random number stay somehow separate within the total. In a sense they do, for if you have the key you can extract the plaintext.

We can generate our false repeats by adding nine more lines of random key to our cryptext, and then taking away 135. As the average of ten lines of cryptext (the nine we have added plus the original) are likely to come to near an average of 135, we can remove the key. In practice we are unlikely to hit on exactly 135, and we don't have the advantage of being able to divide by ten as in the previous example. However, we can repeat the process ten times, getting closer to the average of 135 as the numbers grow, and also we can then divide by ten.

Should you get a message encrypted on an Enigma machine or any similar machine, such as a Lorentz, which was a development of the Enigma, with more rotors and direct input/output via a teletype machine for faster transmission, you can re-encode another nine times on your Enigma; subtract 135 and you will be near the original plaintext. You can then do it again to get closer. Effectively, you will swamp the key.

I will not do a worked example; I will leave that to you, and you can be the first person to break a random number code! You will be famous, and disliked by intelligence agencies around the world, unless, as I suspect, they have already broken it, and weren't telling anyone.

Of course it may not work! There is nothing wrong with being wrong, particularly if you have an interesting journey along the way. But then you have the interesting problem of finding why, and the even more interesting problem of finding a method that will work!

The breaking method is even more powerful than it appears at first sight, for it will also break any scrambling type machine, without knowing how many rotors, or what the settings are, for they all deliver a pseudo random key. For example, if you encode with an Enigma machine, and then delete the original message, you are left with a pseudo random number string, which is just as good as a real one. You could use this as a random string for encrypting in exactly the same way as one from a random number generator, without reference to the machine, and it can be broken in the same way. Just as well they did not spot the method at Bletchley Park, or we might never have had computers!

Should you want an example to play with, here is one. It is from a good source, and even I do not have the key. I may have a go at it later, but will give everyone else a head start!

CHWQG BYPCI AJXKR DGSBS LXFOB VXJFG BIOIL OAPGH NYNGF XTHYQ NBRBK ARPOD NZXDJ

If you break this you will be the first person in the world to break a one-time pad, apart from the security services, that is, who probably do know, but will not let on.

Of course, my method may not work! In which case you can find out why, and will have learned a little more about codes and cyphers.

340

MYTHS AND MODELS FOR INDUSTRIAL ENERGY CONSUMPTION

The debate regarding energy consumption, possible shortages of oil, and global warming with its attendant ills is now starting to mature. For futurists who were well aware of the problem thirty years ago the perception is perhaps of too little, too late; however, the problem of inertia and delayed and inappropriate response can also be modelled, so there is perhaps a lesson for futurists there.

First, a few myths. One sometimes hears that we will 'run out of fossil fuels'. While no doubt some organisations may experience shortfalls due to lack of forward planning, and there may be price hikes as less easily extractable reserves are tapped, there can be absolutely no possibility of running out on the world scale.

The maths and chemistry are quite simple; all of the oxygen in the atmosphere was produced by photosynthesis. For every molecule of O_2 there is an atom of carbon sequestered somewhere in coal, oil or gas reserves; there may be too much coal and not enough oil, but we can make oil from coal, particularly if we have cheap hydrogen on hand. In addition to atmospheric oxygen there is about as much again dissolved in the sea. Also, for the first few hundred million years of oxygen production, there was little or no atmospheric accumulation; the world merely rusted. Even if we only consider the current oxygen availability of a few pounds per square inch on the planet's surface, this still gives an average figure of over a pound per square inch over the entire surface of the planet, seas included, of carbon deposited somewhere. The limit factor for carbon burn is therefore not carbon, but oxygen availability, and herein lies the problem. Obviously we can't burn it all, so the question is, 'How much can we burn?' There must be a figure where further burning gives a negative return; the probability is that we are there already, and will exceed it on the overshoot. Hence the current concern.

Another myth is that there will be an energy shortage. One can of course project one by assuming that demand will double every fifteen years, but this is merely a trend projection, and trends have limits. In the long term more efficient use could actually lead to a reduction in energy use for many industries. In the last century agriculture went from being the greatest producer, to the greatest consumer of energy. With reduction of overuse for fertiliser and pesticides, increased yields and biofuel production, this trend is now reversing. Another example can be found in the comparative consumption of petrol for American and European motorists. They actually spend roughly the same amount as a fraction of income, but the European motorists get a lot more miles out of their gallon.

This explodes the disingenuous statement that raising fuel prices will cost people more. If price rises are phased, over the long-term people will decide how much they wish to spend on fuel, and manufacturers will produce the relevant models. Similarly one sometimes hears that fuel prices will 'hurt the poor.' Usually such statements origin from richer people who are actually interested in protecting their own over-use of fuel. A government can take its taxes where it wishes. A simple solution would be to raise the threshold where income tax becomes payable, leaving the fuel price rise 'revenue neutral', but actually increasing the disposable income of the poor. Who could disagree with that? As a more general consideration, income tax is a blunt instrument; it is far better to tax expenditure, for in this way one can fine-tune an economy.

Another myth bandied about is that you can't store wind energy. This is untrue; you can store hydrogen. This may be inefficient if produced by electrolysis, but remember the adage that 'you cannot waste a surplus'. What is really meant here is that all of a surplus is likely to be wasted, so if you can save only some of it, the normal rules of efficiency of use don't apply.If you go to Dinorwic, in Wales you will find a reversible hydroelectric plant. This was actually built to store surplus electrical production off the grid, which was primarily nuclear, nuclear plants being run at full load as there is little saving in throttling them. This could just as well be used to save power from wind or whatever source. Incidentally it has a start-up time from zero to full production of forty-five seconds, which beats virtually every other type of plant, batteries excepted. The problem in the electricity generating industry is that supply must equal demand with virtually no lag or lead-time. Any kind of energy can be stored, if only by using the surplus in one area to cut use in another.

Compressed air is another form of storage. This would be of great effect if used in conjunction with a gas turbine plant, for nearly half of the energy produced by the turbine is re-used in compressing the air; with no compression load the output would be nearly doubled.

Another myth currently propagated is that the hydrogen economy won't work because you still have to use fossil fuels for hydrogen production. There are of course problems as yet to be solved with hydrogen. It is bulky, it leaks rather well through the smallest hole, and it is explosive in nearly all proportions if mixed with air. On the plus side it is eminently transportable by pipeline and it flows fast with a very high energy density per gram. It is cheaper to transport with less loss than electricity. This means that production facilities can be sited almost anywhere, easing planning and 'not in my back-yard' objections. Personally however, I think there is a good case for a twin track approach, using methanol as a hydrogen carrier for fuel cells, particularly for car transport. There are probably very good rewards in re-examining potential routes to thermal water cracking using atomic energy. I have a distant memory of some work done in Italy, thirty-five years or more ago, using a magnesium-iodine route to crack water at only about five hundred degrees centigrade. There are undoubtedly other routes. Merely heating water dissociates it, as has been discovered by several marine engineers who put too much heat and not enough steam through their superheaters. They glow red-hot and burn out! The explosion that blew the containment lid off Chernobyl and the explosions at Fukishima were hydrogen explosions. The reward will be in a relatively low temperature cycle that strips the oxygen out of steam, and then releases the oxygen, probably at a higher temperature, to recycle the carrier back to strip more hydrogen from more steam. Even Priestly, several hundred years ago, used mercury to first sequester atmospheric oxygen at a low temperature, and then release it at a higher one.

For the quote of the week you can say that hydrogen chemistry is not rocket science!

Another myth is that there isn't a problem with global warming. It is happening, and it will accelerate, for we are going to get overshoot with fossil fuel burning. We should therefore not assume that current and future efforts will solve the problem.

This means that we should also prepare plan B. Apart from the fact that it will be needed, the mere formulation of such planning will serve to point up the problem, and reduce inertia and resistance for those engaged in plan A.

One simple low cost initiative would be to re-map sea levels. Areas of risk could be colour coded, perhaps graded in feet; for example current areas below sea level (or river level for inland situations where increased rainfall is likely to give a similar problem) could be painted red. We could then request all strategically significant organisations within these areas to produce a relocation or flood defence plan and at the same time put a ban on further building or extension of use for such facilities. This would embrace administration centres, major universities and colleges, hospitals, defence installations, trunk communications such as rail and road links, plus electricity grids and major gas and water pipelines. Re-mapping could then extend by contour, up to say the ten-metre line. These contours could then be dated through the next century, so there would be a plan in existence for phased withdrawal with no last minute panics or major infrastructure failures. Further pressure could be brought to bear by consultation with insurers. After all, there is little merit in everybody having to pay increased premiums because a few people seek their own gain by building unwisely in high-risk areas. We could have zoning, where an insurer would only pay out fifty per cent or twenty-five per cent of insured value, with perhaps enhanced premiums for certain areas as well. Planning easement for land acquisition for relocation could be examined.

Another plan B area would be in agriculture. Everybody knows that you can't grow things like avocados, lychees and citrus fruit in southern England. Well, actually you can germinate seeds and they will grow outdoors for a few years until they are cut down by a severe frost. Eventually it is possible that certain varieties will grow here. Bananas actually do grow permanently, but don't crop well. One degree of warming is worth several hundred miles on the map; all growing areas are going to migrate towards the poles. Thus there would be a case for starting an 'experimental corridor' where plants are grown further north than their normal range, to get early results on what can and cannot be achieved. Agricultural institutes, schools and colleges could be grant aided to perform a very large number of simple experiments over a wide area. The results could be posted to a collection and evaluation site. This would form a database for farmers later; they would know what could be grown in their area, as current crops became less productive. There could be student assistance for young agriculturists to spend time a few hundred miles further south to see how farming methods vary with latitude. We could do another set of maps with projected isothermals, and perhaps less accurately, rainfall.

This kind of planning is necessary, cheap to implement and gives a positive action profile for the general public who need to be involved politically because they have votes. Emergency decrees from on high are not a good way to govern, if only because they are likely to be ignored or subverted if the public and industry does not know what is going on.

Greater economy of use for energy is of course always worthwhile. Surprisingly, the accepted wisdom that compression ignition engines can only convert about forty per cent of the energy in the fuel to work, with spark ignition at about thirty per cent, is not true. A relatively simple redesign to a somewhat modified cycle can increase yields to about sixty and forty-five per cent. The extra cost will soon be returned in savings as fuel prices rise. Details are elsewhere in the book.

Another area where efficiency can be enhanced is in weight saving. Generally the greater the technology, the less the weight. Planes weigh less than ships, but perform a similar function. A modern glass and steel office building weighs less than the reinforced concrete and brick infill of yesteryear. Similarly, we don't need a car weighing nearly a ton and burning over a ton of fuel a year to carry one person around. People still use such behemoths to go and buy a loaf of bread. The starter motor provides more power than is necessary for this job. There would be a case for a taxation class system where weight and energy efficiency is part of the equation; after all it is principally the weight of a vehicle that causes road damage. Also, there is less energy cost in production if less material is used. Thus development of composites, plastic and cellular construction methods should be fast tracked and embraced as part of fuel economy measures.

Another area where efficiency could be increased is in evaluation of what actually goes on in decision making. This can be sensitive and controversial, but we also need to address sensitive and controversial issues. We have all had experience of doing really good work, to see it sidelined by vested interest, bureaucracy or inertia.

A friend of mine once got a first job out of University as a progress chaser for a major organisation. His job was to see that everything about contracts was on track, chase under-performance from subcontractors, use minimal path routes, reschedule where necessary, bring pressure to bear, and generally keep the show on the road. He approached the job with energy and enthusiasm. This was slightly dampened by the end of the first week when he realised that quite a lot of his energy and enthusiasm was spent in dealing with anti-progress chasers, their brief being to buy time, erode deadlines and generally minimise his impact.

Another girl I knew, having just got her chemistry degree, was thrilled to get a job with a toothpaste manufacturer; work at the cutting edge of preventing tooth decay. Actually her first task was to find a way of putting stripes into toothpaste. I would have thought a good lab technician would have been more suitable.

There is a perceived problem amongst environmentalists concerned with energy usage, about congressional and parliamentary lobbying on their respective sides of the Atlantic. Their own lobbying is perceived as perfectly above board, of course.

Now, lobbying is perfectly respectable, within guidelines. A situation can arise however, where the short term and legitimate concerns of an industry can really be against the equally legitimate but long term concerns of national interest. They can be seen as punching above their weight and having undue influence in representing what is really a minority interest. There is here a case for the long-term national interest taking priority. We have seen this in the case of subsidies; perhaps the European farm subsidies are an example. Not that I wish to make this a particular case; there are plenty of others. There is the perception of an industry being propped up at increasing cost to the public purse until it becomes unsustainable and there is a sudden collapse leaving a lot of people stranded, whereas a gradual shift of production and emphasis would have been much more productive.

There are national agencies that have a brief to examine threats and defend the interests of a country. Generally threats are considered as being posed by evil foreigners, whereas some of the worst ones may be posed internally by not at all evil people, pursuing agendas that while perfectly legitimate in themselves, don't jibe with the best long term interests of all. Without mentioning specific industries or agencies, it might be worth extending the brief of defence agencies to embrace a long-term overview of internal trends and decision making, highlighting where there may be undue influence and exploring the mechanisms by which this comes about. This need not be perceived as a threat, or confrontational, but merely a problem in conflict of interest resolution. Everybody could get on with their particular area of expertise as long as they knew the game plan. Such agencies are particularly well placed for this, as they have evidence gathering experience and are in a position to assess international perception, which impinges on the reception of initiatives from developed countries directed towards the third world, itself a major area of concern in fuel policy.

To chalk up some examples of common misperceptions, everybody in the UK knows that American cars have awful miles per gallon performance figures. A lot of these worthy citizens are completely unaware that the American gallon is twenty per cent less than the British one, while the continentals don't know how many litres there are in a gallon of either variety, or how many kilometres there are in a mile. If it is that bad close to home, no wonder that perfectly reasonable proposals can sometimes be viewed with less than the anticipated enthusiasm further afield.

Since there are vested interests engaged in decrying climate change, it seems to me to be legitimate to state the opposite case: There will be climate change; this will lead to most capital cities and ports being flooded, as they are close to sea level. There will be considerable loss of cropland, loss of forest, and huge population movements. There is the probability of wars over water rights and mass movement due to desertification, flooding and forest fires.

Major epidemic outbreaks will occur as health services break down leading to a major reduction in population, which will in part reduce the problem. Instability will reign and the average per-capita income will drop in all countries. And all this will be the fault of the climate deniers!

ON NANOTECHNOLOGY

I dabble in chemistry occasionally. I am an armchair chemist now, as I have an armchair but no lab; also I am a great believer in Gedanken experiments. Not that I was all that good when I had a lab, being left-handed with my left hand and cackhanded in both; on a bad day it can take me half an hour to wire a three pin plug. Poor innervation or something.

I have had a few little successes however. When I was at college in Portsmouth we had a set of Geissler tubes, and a high-tension induction coil to fire them up to show light production from inert gas. I think it must have dated from the early part of the 20th century, when neon lights were just coming into vogue.

At that time we were taught that the inert gases were inert, because the outer electron shell was filled with eight electrons apart from two for helium, and they could never form compounds. I did not believe it. After all, if you could ionise them by knocking an electron off, why not present the ion with an electron from somewhere else to see if it would accept the bait? The obvious bait was fluorine, as it only had one electron in its outer shell and was highly reactive.

This presented a problem. Fluorine was unobtainable as a laboratory gas then, in the mid-fifties, all production going to isotope separation, and it is not that easy to make, being worse than chlorine. Nothing daunted, I decided to use hydrofluoric acid, making a lead crucible by melting some lead into an old tobacco tin and dishing it with a ball headed hammer. Fluorospar provided hydrofluoric acid fumes when heated with sulphuric acid. Argon was obtained by breaking dud lightbulbs underwater and collecting the bubbles; you need quite a few as it is at very low pressure.

We had a reasonable spectroscope, and the idea was to look for strange lines in the spectrum, to see if we had a new compound, the idea being that there would be a displacement after we had given it a blast from our high-tension coil.

So did we find any new lines in this long shot experiment? Actually, we found lots, for us, that is. Far too many in fact! Have you any idea just how many spectral lines there are? There are thousands and thousands. Even iron has two thousand. To start with we had half the elements under the sun. Our lead had antimony bismuth and arsenic in it, plus probably a few more trace elements. There was the double yellow of sodium through all. Actually I did not mind that much; it acted as a handy reference for calibration, and after a while you don't notice the dominant lines It is a bit like weeding the lawn, where you get tunnel vision on weeds of a certain size and miss the really big ones. In the midst of this plethora, I read in some old journal, I don't remember which, but possibly the transactions of the Chemical Society, that someone may have made inert gas compounds in the nineteen thirties, so I rather lost interest and went on to other things.

I did not publish, as I believe in private circulation papers. Also it is a bit difficult to publish controversial findings; try it someday. Personally, I think publication of experimental results is hopelessly slewed. Everybody wants to have good positive results. It is just as important to report the negative ones, as we are now beginning to realise regarding clinical trials from Big Pharma. Even more important are the inconclusive ones that are worth of further study. Really good scientists seem to be able to slip them in, or used to. Cavendish, as early as 1785, reported that $1/120^{th}$ of the initial volume of air was left over after removing all known gases; enter the rare gasses. Well, not quite; it was nearly a hundred years later that Rayleigh suggested that there was another gas.

Millikan, working on the charge on the electron, using fine droplets balanced between an electrostatic charge and gravity, noted that he occasionally had a droplet that appeared to have a one third charge. Strange, as we now know the electron has its charge made up of three equal parts, although I don't see how they could become separated on an oil drop. Michelson and Morley got some slightly anomalous results when finding the speed of light, but I don't think this was followed up. In the very early days of radio, they got a strong reflected signal three minutes after sending. This is quite unexplained, unless you believe there is a transponder sixteen and three-quarter million miles out there somewhere.

If you want to check, it is impossible with all the radio clutter about now. However, if you placed a transmitter on the back side of the moon, you could try sending signals to see if you get the echo. A small satellite could pick them up, and deliver them back to earth. A low orbit might be best, as things go slower around the moon; ten miles would do, and cost less to get it there.

Sometimes I did publish. I suggested the use of the rare earths, particularly lanthanum, as a candidate for superconductivity, in a none too serious article entitled 'Atomic Allsorts', for 'The Inventor', the magazine of the Institute of Patentees and Inventors, a few years before the seminal paper regarding their use. Not that I knew much about superconductivity, nobody did then, but I was merely taking a lucky guess about the rather peculiar cometary electron configuration the transition elements have; some can theoretically go through the nucleus, building the penultimate shell. The magazine is now called 'Future and the Inventor' and I still publish occasionally. In the same article I suggested the possibility of an alternative route to atomic energy, by stripping electrons off an atom. After all, a free neutron has a half-life of only twelve minutes or so, so why not see if we can get them to decay in, or at least be ejected from, a stable isotope? It was some years before they caught up with that one. An isotope of technetium is now used as a power source for satellites; irradiate it with hard X-rays and it undergoes transformation. Switch off the X-rays and it is inert and hardly radioactive, very safe if your rocket has a launch failure. I also mentioned the possibility of trying to get isotopes to swop neutrons. If we take the 'ideal' atom as having an equal number of neutrons and protons, most of them don't; there are plenty of isotopes with too many neutrons, put in more still and they become unstable. My idea was that isotopes are potentially less stable the more spare neutrons they have. So perhaps we could do a minor rearrangement, a sort of low-key atomic reaction. The packing fraction would be less, of course, rather like if you go on a weekend you only lose your toothbrush, while in a week you can lose a whole pair of boots, or at least one of them. The lower energy yield would still be a lot, while there would not be such a mess of by-products. When Pons and Fleishman proposed cold fusion, everybody was looking for free neutrons, and said it did not work because they did not find any. If you could get neutron exchange, there would be none. Actually, you don't even need an exchange, just an electron swop between protons and neutrons, and a bit of energy. Palladium has a couple of electrons with a sort of cometary orbit that penetrate right through the other shell; statistically they go through the nucleus too, so there is a sort of clue there. There are lots of isotopes of palladium to try with.

All this of course is woolly science. This is deliberate. At one time religion was all about certainty, science about doubt. It seems they have now changed places somewhat. If you adopt as your motto that 'It is necessary to believe all things possible, in order that some may become so', you are halfway there. Forget about being wrong; chuck a whole mismatch of ideas into the omelette and see what comes out. If you start with a lot of unquantified possibilities, they can be sorted into improbables and more probables, and eventually you may get something that works, even though the received wisdom says it can't.

UNIVERSAL STANDARD TIME: FASTER THAN LIGHT, AND A LETTER LATE AND EARLY

Let us suppose that it is your birthday, and a friend has brought you a pair of gloves as a present. However, her ten-year-old son has insisted on wrapping them separately, on the theory that two presents are better than one. Unfortunately she has left one package on the train, probably under a discarded paper, for she rummaged in her bag for a pen to do the crossword.

You make light of it, saying, "I always have my walking stick in my right hand, and put the left in my pocket if it is cold. Let us see which we have!" You open the package and of course it is the left one. Now, if you think about it, it is rather amazing that you have instant information about the contents of other package, without ever seeing it, or knowing where it is. There is no time of transit for information, no address, and no doubt.

Looking at the situation in more detail, the reason is that a pair of gloves is not two separate items, although they may be separated, but one thing. The separation in distance is a bit like a magician's diversion of attention; it is not relevant. We could say that the existence of a left glove of a pair implies the existence of a right glove. In a sense they are implicate. Until you open one package, there is a fifty-fifty probability you will find a left one. Once you have opened it, you know, and the probability is crystallised as a one way certainty. Of course, for a macroscopic object like a glove, it always was a left one, it was a left one yesterday, and will be tomorrow. Were a glove sentient it would have known it was a left one, and it would have been so at the other end of the universe, apart from the logistical problem of getting it there.

353

You will of course by now have gathered that the discussion is not about gloves, but implicate quantum particles.

It is possible to create two items, photons, or maybe neutrons or electrons, paired and implicate in a similar way to the pair of gloves, with a different handedness, usually spin up, or spin down. They are not interchangeable, or at least, not in three dimensions. A left handed corkscrew can never become a right handed one, unless you can turn it through a fourth dimension. In the same way, the spin of a particle is related to the field of the particle, either one way, or the other.

But not until you look at it, that is. Unlike the gloves, according to the Copenhagen interpretation, which is one of several, the spin state is a mixture of the two, superimposed and not expressed until the collapse of the probability state forces it to be one or the other. Unlike the glove, the photon does not know what spin it has. This is not a gap in our knowledge, as with the gloves, for there is nothing to know.

This is a little more interesting. We can fix the state of distant photons, immediately, and without leaving home. Actually photons are not good candidates; they are bit ephemeral and flip at the slightest disturbance. Electrons are much more stable, but for the argument, it does not matter much which we use.

This gives at least the possibility of faster than light communication. It is said that nothing can travel faster than light. Actually no 'thing' can, for to accelerate it requires energy, and energy has mass. Eventually, as you pump more energy into a particle to try and get it to speed up, you are merely making it heavier. Light itself seems to have a fixed top speed in a vacuum. Quite why it is that particular speed is anybody's guess. It could probably tell us more about the nature of the universe than the speed of light, if we knew the right questions to ask. Also of course it may go slower in different media; can be stopped in fact, and stored for a while. If you ask a photon what speed it travels at, it won't understand the question. It does not inhabit our time, but the everlasting now. Its speed is infinite by its clock, which has stopped, or it does not travel at all, depending on how you look at it. For the photon, the universe is flat.

However, information is not a 'thing'; it has no mass, no energy. Write the whole secret of the universe on a sheet of paper, and it will weigh no more than the paper and pencil did separately at the beginning.

There are a few problems of course. To start with, you have to get your implicate quantum particles to their far destination, at the speed of a bicycle, or rocket, which is much the same thing when compared to the speed of light.

Do not despair though. You can make a phone call across America in a split second, but the telephone line took months to set up, mostly travelling by horse and cart. Similarly, you could send a packet of implicate photons every day, or every month, on a regular basis, so that there was always a supply at the other end. The same thing applies when you turn on your tap. Water comes out almost immediately, but the reservoir is a hundred miles away. The water of course does not travel at more than a few feet per second; it is a different lot of water you get out of your end from the bit that goes in at the other end, but it does not matter. You can do the same sort of thing with a radio signal through a crystal, over a short distance. The time of transit through the crystal is less than through a vacuum.

There could be very big advantage in faster than light communication. Even a telephone conversation to the moon is a bit stilted, with about a three second delay. To Mars it would be more than tedious; you would have forgotten the question before you got the answer. If we could have instant communication, we could drive a rover from our desk, rather than try to design intelligent ones that will get stuck and pass by the interesting bits. You could even get information out of a black hole, assuming there is any in there, that is. It may have been destroyed, which breaks the second law of thermodynamics, or be stuck on the event horizon.

The trouble is, of course, we do not know until we pop the question what the answer will be; a series of random left and right hand spins counterpoint at each end is not much help. We can actually send information by doing it statistically, by using a large number of photons for each bit of information, but I won't go into that now. What we may be able to do is to 'force the spin'.

For example, we could, while examining our half of the pair at our end, open the box in a strong magnetic field, so that the answer was always spin up. We would know then that the other one of the pair would be spin down. By switching the field we could send a signal in binary code. It would take a while to get our supply of implicate pairs in place, but once in place we would have our link set up, rather like a telephone line that transmits far faster than its rate of erection.

So there we have it. Faster than light communication. Not just a bit faster either; the communication is instant. Just as an individual photon does not experience time, our system is time-proof. The secret is, just as with the gloves, that the paired photons are one thing, not two separate items, although they may be separated. The separation is a diversion and not relevant.

This leads to another interesting effect. In mapping the heavens, just as in mapping the ground, a baseline is useful. Unfortunately the biggest one we have is that formed by two points at 180 degrees around the earth's orbit. This is just about good enough for parallax measurements to the nearer stars not in the same plane as the orbit. An observatory on Pluto would be more useful, apart from its rather languid rate of travel, that is. But why stop there? We can send a probe way out of the solar system, giving us an ever-increasing baseline that would give increasingly accurate parallax measurements, to more and more distant stars. While we are at it, we might as well send two in opposite directions, to double the baseline; four would be better still, arranged at the vertices of an equilateral pyramid, to cover the other two planes, tilted so we don't lose one behind the sun.

A high degree of accuracy would be needed however in knowing where our probes were. A small discrepancy in position would be magnified enormously at the very small angular displacement we will be looking for. We begin to run into problems of time. More specifically, whose time do we use? For the clocks on the probes would not keep our time. To start with they are moving, and their speed is not constant relative to us or to each other; also they are in a much weaker and declining gravitational field. What we need is a clock that will give the same time at all four points.

Enter implicate particles. If we have a supply of particles on board our probes, we can be sure that nine-o-clock Tuesday morning on board will correspond to precisely nine-o-clock back home. We have in effect a universal standard time, just as Newton supposed, and different from Einstein's variable rate.

This lack of time has another potential effect. When I was six or seven, I wrote a letter to myself, to be opened in the far future, maybe when I was ten or eleven. I was greatly taken with the idea of writing a letter into the future. I could hardly wait to get it! Actually I waited three days, and when I opened it I knew all that anyway and it was rather an anti-climax. What would be more interesting would be to write a letter to yourself back in the past. More useful too, for you could include the winner of the Grand National and the stock prices.

In fact, one cannot actually communicate with the past or the future. All action takes place in the present. If you write a letter to someone now, in the present, they will get it a few days later, in their then present. If we could step outside time then any bit of any present could communicate with any other bit, just like our implicate particles.

For example, if we can force the spin of one of a pair, the other will always have the opposite spin. It will also have it yesterday and tomorrow, in our time, for its time is all of time. Thus if we set up an implicate pair, leave them on the lab bench for a day, and then force the spin of one of them, we will know what the spin of the other is. With luck, we could open the unforced one first, and of course find which way the spin of the other was forced tomorrow. Thus again, we have a way of communication by binary code. A collection of pairs could be used to spell out the name of a race winner. The information is always in the present, for implicate particles only live there, but we have connected two bits of the present across a gap of a day in our clock time.

The only limit is that we could not go back beyond the date at which we set up our link.

There is just a slight problem in that by opening one first, we force the other; we may well trip its state into being, thus ruining the experiment. No problem. Enter virtual photons. These have the advantage of not really existing, and so go unnoticed by our captive implicate particle. If you have ever wondered how a beam of light knows how to take the shortest route, it does so by sending out virtual photons over every route, and working out the best path by a series of averages. It does this in no time at all, and the real photon makes its leisurely way acting on this information. The theory is called Quantum Electrodynamics, and it is the most accurate theory we have. Of course things may not really happen like that at all; we could be looking at a projection of some deeper reality, but it is our way of explaining it and it works. A variant, Quantum Chromodynamics, works for forces acting on particles within the nucleus, although gravity is resistant at present. Thus we may be able to spy on the contents of the box without actually opening it. Send a virtual photon in, and it will exit if the real one is spin up, for example, and not make the transit if spin down.

There might be a temptation to try to trick the experiment by deliberately choosing the wrong spin tomorrow, having got our results yesterday, in which case you would have left the Copenhagen interpretation and moved into one of 'Many Worlds', another parallel theory to the Copenhagen one. You could still win your bet, but another you in another world would collect the money. It would probably be best to maintain a complete separation between the input and the readout, with two different people, not in communication, performing the two different functions.

So we may be able to find out the content of the box by seeing what a particle that does not exist does not do. Now that is a real magician's trick. It would be death to gambling! Well actually not; there is a defence strategy, which I will not detail. If you are in the business you should hire a good consultant.

Of course the science fiction writers have done it all before. Ballard explored time, while the first mention of Windows, the computer program we all love to hate, was in a science fiction story, 'Now is Forever', by Allen, as a throwaway line; a little recursive graffiti in the halls of time. The story is about a computer and time, or the lack of it, published by New English Library in a collection called, coincidentally or not, *Stopwatch* (ISBN 45002142 4), page 139, line 7 in the hardback edition. The interesting bit is that it was published in 1974, which is a bit before Windows was invented. It looks a bit recursive to me.

Most have heard the old proverb, 'A stitch in time saves nine'. You may have wondered how one could stitch time? Well, now you know!

GETTING IN A SPIN

All stars spin and neutron stars spin like fury, more so than you might think, as there is considerable time dilation from our point of view. It is therefore very likely that all black holes spin, and this presents a problem. The infalling matter gains a very large amount of kinetic energy, and this adds to the spin. As some people don't seem to believe this, I will explain.

Let us imagine we are in a children's playground, where there is a carousel or merry-go-round, which you can spin up by running at the circumference and jumping on tangentially. This carousel is slightly different, for convenience of the experiment, in that it consists of an aluminium beam, with a handrail down one side, rather like a gangplank as can be found on small ships. Also, as it is an imaginary one, we can think of it as having infinitesimal mass.

For the first experiment you run and jump on; your speed does not change, but you follow a circular path, giving rise to what is commonly, if incorrectly, called a centrifugal force, proportionate to the radius and speed.

For the second experiment you run and jump on again, but this time at half radius. Your speed is the same, but the circle is only half as big, leading to double the rotation rate and centrifugal force. The total energy is the same in both cases.

For the third experiment you jump on as in the first experiment, at the rim, but this time you pull yourself along the plank until you are at half radius. To do this you have to do work against the centrifugal force. So where does this work go? It must have been stored in the spin of the carousel. Not only are you spinning faster because of the redistribution of mass, but also there is an extra component due to the work you have done. This extra spin also adds to the centrifugal force, requiring progressively more work as one proceeds down the gangplank, and it would be impossible to achieve the centre.

For stellar collapse, although some of the work is stored in nucleosynthesis for atoms heavier than iron, and some blown off, this is only a fraction of the whole. As the contraction proceeds, the spin tends to infinity, and so also does the centrifugal force. The black hole formed is unlikely to be a sphere, but a torus or ring. There would still be collapse in the plane at right angles to the spin, so instead of collapsing to a point it would collapse to something akin to a string. There might be some energy radiated, slowing the spin, but not past the event horizon.

There is an added problem. As the kinetic energy mounts during the collapse, so also does the mass of the string, for energy is mass. Presumably this mass will be robbed from the centre, the two masses tending to parity, for the total mass must remain the same. In this scenario, the mass of an object is partly in the gravitational field around it. Interestingly, this may mean that the mass of an atom is not a constant, but will decrease the nearer it is to the centre of a galaxy, by virtue of the greater amount of kinetic energy destroyed in its fall there. We have a somewhat similar effect in magnetism. If I use a cryogenic electro magnet to do work, then the circulating current must be robbed by that amount of work done. If I do work against the magnet, the current is enhanced.

Similarly, if I pick up a brick and put it on a table, I have done work against gravity, and the mass of the world and the brick must have increased by the mass equivalent of that amount of work done, but my store of available work, in the form of chemical energy in food, reduced. If I push the brick off the table, work is done by gravity and dissipated as heat on impact, leaving the system, the world and the brick growing correspondingly lighter again.

THE INFORMATION REVOLUTION

When I was six, during the war, we had a shipwrecked family stay next door to us for a short while. The mother and five-year-old daughter had been on a ship from America, to join the husband in the UK, but it had been torpedoed. All their worldly goods were in two small wicker laundry baskets.

Strangely, amid the salvage was a treasured toy, a miniature wind-up gramophone There was only one record; light brown, cracked and about five or six inches across. We played 'Little Miss Muffet', and 'Mary had a little lamb' endlessly under the stairs.

I thought it was marvellous! We did not have a gramophone, and a recorded voice was much more of a miracle than the radio. Later, a boy at school had a wind up full size portable of thirties vintage, and I had the first Phillips stereo in the sixties. Thirty years later I gave it to the national radio museum, along with an original wax cylinder Edison and a sixties radiogram I had collected along the way.

It is strange to think how quickly things become obsolete nowadays. The whole history of recording relegated to the museum in one lifetime. I used to get hundreds of discs free through the post; Tiscali, AOL, the human genome, antivirus programs, catalogues and so on. Each one can hold more than half a ton of cracked brown records. On the way millions of man-hours and billions of dollars have been spent on the technology and programming, while progress accelerates. Within a few decades we will have quantum computing, millions of times faster than the best we have now. This may slow things down even more; someone once did a study to see what effect inventions had on society. Blast furnace steel? A huge expansion in manufacture. Railways? A vast increase in trade. Computers? Not a lot, except everybody started playing games. The mobile phone has actually decreased productivity on most building sites; you only have to observe how often workers are immobile, head slightly in the air, shouting down the phone. I suspect it is the same in all industries.

Also, availability devalues. I have never put any of my free discs in my computer. Instead my wife has hung them over the soft fruit patch, where they flash their bright spectrum across the garden.

I expect they will attract the birds from miles away.

A TWIN PARADOX

After a successful run of fifty, a TV game show has a final Christmas edition in which some of the most successful items are re-jigged.

The first act is where there are three identical boxes containing a one-penny piece, a two-penny piece, or a fifty-pound note. A contestant chooses a box but is asked if they want to exchange it, after one is opened by the host.

We join the show as the host draws a name from a hat, previously passed round the audience.

"And here we have the lucky winner, Eve Eva! A big hand everyone!"

However, not one but two girls start to make their way up the steps at the side of the stage. One is detained briefly by someone in the wings, a few words are exchanged, but she continues on to join the other on stage.

Host: "It seems I am the one with the puzzle; not Eve Eva, but Eve and Eva. Identical twins! However here is the producer with an extra chair, so it must be all right!"

"We always do things together!" exclaim the twins together. "Sometimes we think of ourselves as the same person, twice over!"

"Well, I hope you choose the same box," exclaims the host. "Otherwise I will have double trouble."

"Box number two!" they say, both together. Then, pre-empting the question, "Because she chose it!" each indicating the other.

"I will put it on this table in front of you," says the host, so doing. "And I will now open one of the remaining boxes." He opens number three and holds up a two-penny piece.

"Now, Eve," he continues. "Do you wish to exchange your box

number two for the remaining box, number one?"

"But of course!" replies Eve.

"Would you tell me why?" he asks.

"When I chose my box," explains Eve, "I knew there was only a one in three chance of it holding the fifty pound note. This is still true, for opening a box cannot alter what is in any other one. However, now one box that did not contain it has been removed, there must be a two in three chance of it being in the one I did not choose. The amount of the prize does not make any difference, other than as an incentive. They could just as easily contain A, B, & C, with no value differential. The probability is only statistical, of course, but if we played a large number of games the reality would settle down to two in three for my second choice."

"I cannot fault your answer," says the host. "Now, Eva, do you also wish to exchange the box?"

"Of course I do," she explains. "For the same reason."

"Another perfect answer!" crows the host. "And what will you do with the fifty pounds?"

"Oh, but I don't want the fifty pounds: I want the penny!" exclaims Eva. "As I came on stage the producer stopped me, and said if I won the penny, he would exchange it for two tickets to the grand charity fund-raising party next week, as he wants to auction it to see how much he can sell one penny for. We get all travel and hotel expenses, present the penny to the lucky bidder, and meet all the celebrities!"

"But your sister has a two in three chance of winning the fifty pounds," wails the host, counting on his fingers, "You can't both be right!"

"What if you had asked me first?" asked Eva. "Would that have made any difference? Anyway, you said I had a perfect answer!"

"Now I do have a problem!" says the host. "But here is a note from the producer; he got me into this, perhaps he can get me out!"

An envelope is handed over and opened.

"Oh," he says. "You have both won; two tickets and a fifty pound note and a penny. There are also two more tickets for the first lucky viewer at home to send in the correct solution, I won't open the box until the lines close at the end of the show."

He tips their box so the bottom faces the camera, displaying an e-mail address and telephone no.

"Best of luck, and on with the show!"

CROSSTALK:

INDUCED TRANSITORY CURRENT FLUCTUATIONS CAUSED BY MOVING A CIRCUIT IN A STATIC ELECTROMAGNETIC DIFFRACTION FIELD

In the early '40s I read an aviation story, written in the '30s, in a 'Boys Own' type of adventure storybook. The subject was that of a test pilot who was instructed to fly a set circuit and report if anything unusual happened. The result was that at a couple of points in the circuit, his engine cut out but could be restarted after gliding and losing height.

The point of the story was the testing of a 'secret ray' that could cut out an enemy aero engine. There were no technical details given, but of course there were early experiments in radiolocation going on at that time, and the author may have got wind of that.

In the mid-fifties, whilst travelling through the New Forest on a 1936 BSA Empire Star motorcycle, my engine cut out whilst passing a few dozen yards away from a small military camp sporting a lot of radio aerials. Exactly the same thing happened, in exactly the same place a few weeks later. I remembered the story and wondered if there was a connection.

Now, a radio transmitter will induce a fluctuating current in a wire. Also, in a wire already carrying a current it will introduce fluctuations in that current, the induced current being imposed on the existing current to raise or lower its value according to the phase of either or both. It would be possible to phase match so that the current in the wire was annulled by the imposed signal current.

However, in common with many motorcycles of that era, the ignition was a dedicated circuit fed from a magneto, being entirely separate from the lighting circuit. The only piece of wire exposed was about a foot of HT lead, the coil winding being enclosed by the magneto casing in what was effectively a Faraday Cage. Also, the HT current was of very high voltage, unlikely to be materially affected by the weak induced current. Thus I considered such an effect unlikely, although the two events probably had the same unexplained cause. However, Faraday cage or not, the whole of the metal of the bike was insulated from the ground by the tyres, so the 'earth' of the circuit was in fact a false earth that could fluctuate in value with regard to the protected bit inside the magneto casing.

That such effects do exist is well known; my science master said he knew of someone who lived more or less under a BBC transmitter in London, who ran some lighting off coils in his loft! In the early sixties I noticed that when passing in front of the big early warning dishes on top of Portsdown Hill on a foggy day, there was often an absence of fog in front of them.

Professor Laithwaite told me once that when demonstrating a flexible linear motor at Imperial College, a steel ball bearing was used to rotate in a linear motor bent in a vertical circle. After switching off, on one occasion the ball started to lose speed, breaking contact at the top of its loop and hopping across the circle, but then speeded up again. I suggested there must have been a source of electromagnetic radiation somewhere at hand, which was being picked up by the motor windings.

Once, on a boat up the Thames, moored near to an aerial array, we found that the boat lighting was flickering at the rather unusual frequency of 43 cycles per minute, actually a brainwave frequency. We moved the mooring.

Now, where there are two or more transmitters, a stationary interference pattern will be set up between them, consisting of nodes and antinodes where there is a sharp fluctuation in field value. The spacing of such nodes will vary as to the distance of each node from each aerial. By drawing concentric circles from two different centres, the variation of overlap pattern can be readily seen; the same effect can be produced in a ripple tank, or a similar diffraction pattern can be produced with light, as in the two slit experiment. One of the early navigation systems used the same principle. Flying a test circuit through such a diffraction pattern will give rise to a fluctuating current in the circuit. The rate of fluctuation will be related only indirectly to the transmitter frequency, being a function of the pattern itself, which will vary over distance, and the speed and direction of transit. For example, if nodes are spaced at one meter, and an aircraft flies through the pattern at a rate of 400 mps, then an oscillation of 400 cps will be produced.

In aircraft circuit design, 'crosstalk' between different circuits by way of induced currents is guarded against, and tested for. However, it is unlikely that such a testing regime would be proof against an exceptional transient external effect such as the aircraft flying through a strong field produced by, say, two radar transmitters giving rise to an interference field only distantly related to either of the fields separately. Such an effect might give rise to a sufficient temporary overload to trigger a protection circuit shutdown.

Thus one element to be looked for in the case of unexplained shutdown could be the production of a fluctuating field occasioned by the movement of a circuit in a stationary diffraction pattern set up by two or more transmitters.

SECTION 8

MAGNETOCHEMISTRY

Again, a mixed bag of bits and pieces. I hope some of it is interesting to some people.

Chemistry has been around for many thousands of years; metals were extracted, pigments and dyes prepared and food processed long before Democritas wrote his atomic theory, borrowing from earlier Greek work.

However, it was not really placed on a quantifiable modern scientific basis until the arrival of the electron theory, in the latter part of the nineteenth century. With this, all reactions could be explained and predictions as to reactivity made. This led to a new field, electrochemistry, where an electric current was used to force difficult or otherwise impossible chemical reactions, the most commercially important being the production of aluminium metal from its oxide. It could be said that all of chemistry is nothing but the arrangement of electrons and their associated fields in relation to each other and the nuclear charge.

There is some mileage left here; for example, it would be interesting to see how chemical reactions are accelerated or retarded in the presence of a deficit or superabundance of electrons. It is possible that an electrostatic charge could act as a catalyst, or to move the equilibrium point and therefore the yield and rate of a reaction, for many reactions do not go to completion: as the product increases and the feedstock decreases, back reactions occur and the rate stalls.

There is also the possibility of disrupting the nucleus by peeling away successive layers of electrons, particularly for isotopes, which are only metastable. There might be a case for exploring the possibility of low yield atomic energy, not by fission or fusion, but by adding or subtracting a neutron from common, otherwise stable and abundant isotopes. Chlorine springs to mind, but there are many others. This process is to a degree used in medical and tracer isotopes, which are naturally radioactive, and in power packs of satellites, where an isotope of technetium, normally stable and unreactive, breaks down when irradiated by hard X-rays, the photons of which disrupt the electron shells. When other workers examined Pons and Fleishman's claims to low temperature hydrogen fusion, nobody looked at the isotopic ratio of the palladium they were using; there are half a dozen or more in varying concentration according to the source. It would be quite possible to swap neutrons between these without giving rise to detectable free neutrons. This would release the energy differential between such isotopes,

One problem with removing electrons from atoms is that they are likely to take them back again in short order, from whatever happens to be around, and the closer to the nucleus the more energy is required, transition elements excepted. High electrostatic charges are prone to lightnings! This need not be much of a problem if 'nested containment' were used. For example; if you were given an air pump capable of generating a pressure of a couple of pounds per square inch, plus a plastic bag, and told to achieve a pressure of two Ats, the job is impossible. However if you are told you can have as many air pumps and plastic bags as you like, it is simple. All you have to do is to put one bag and pump inside another, until at number eight, you have your required overpressure. The same idea could be used to contain more and more energetic ions created by stripping atoms. You might find you have created a unique brand of particle accelerator in the process, so the idea might need tweaking.

However, this by way of example as a possible use of an electrostatic field. What I really wish to explore is the use of magnetic fields to influence chemical reactions.

As a thought experiment, let us suppose that we have a powerful horseshoe magnet, in between the poles of which we introduce a piece of iron wire. The wire will of course be drawn in until an equal amount pokes out on either side. Further, we can make the magnet do work while the wire is being drawn in. Of course, to get it out again we will have to give back the work we extracted, but we don't have to give it back in the same form.

It is pertinent to ask where this energy comes from. Simply, it comes from the electrons in circuit in the magnet. For an electro magnet the current will be robbed of a little energy. For a permanent magnet they will lose energy and contract their domains slightly, and the magnet will be somewhat less capable of doing further work by virtue of that extracted. It should contract a little, and this could be measured. The effect is more noticeable in piezoelectric crystals, such as quartz. Compress the crystal, and you get a little electric charge out, supply electricity, and the crystal expands again. Even a common sugar crystal will emit a photon if crushed, although rather dim.

Let us now reintroduce our wire in between the poles of our magnet, just past the centre point; we allow it to dissolve in sulphuric acid, yielding the less magnetic iron sulphate. Now we can get much more work, feeding in wire until all the acid has reacted. We might think that we could run down the magnet completely by this process. However, it is likely that the chemical reaction would re-supply the energy in just the same way as if we pulled the wire right through, thus putting back in the energy we got out as it went in.

Let us suppose that we could ramp up the field of a superconducting magnet indefinitely. We could extract more and more work as the field increased, but the reaction would slow and eventually stop as it had to do more and more work to counteract that which we took out as the wire fed into the field. Eventually the reaction would stop once there was insufficient energy available to pull the iron atoms out of the field. Increase the field further, and pull harder on the feed end of the wire, and we could reverse the chemical reaction; iron atoms would be torn from their sulphate bond and rejoin the wire.

Of course this reaction is rather energetic, and it might be a good idea to start with something where the energy requirements are a few orders of magnitude lower. We could for example use a much weaker acid, or possibly achieve a separation between iron and nickel, which are often mined as mixed ore and are difficult to separate due to their chemical and energetic similarity.

We are not only restricted to magnetic materials or conductors. All atoms show paramagnetic or diamagnetic properties, according to which way they line up in a field, and a very high field would alter the orientation of molecules. This could be useful in depositing thin films of atoms as are used in transistors. We might be able to increase the rate of deposition, or its accuracy, or maybe deposit atoms that cannot easily be lined up otherwise.

For a chemical reaction to occur, apart from the requirement for a favourable energy gradient, the atoms or molecules have to be in the right place and in the right orientation. Random thermal motion will see to this, but for large molecules the requirement slows the reaction. If the components of a large molecule could be presented the right way round and in the right place the reaction would be speeded up. A rotating or oscillating magnetic field might help; a combination of magnetic and electrostatic fields might be better. 'Pumping' could be used to reinforce a desired frequency of oscillation, rather as a laser is powered. This might enhance the reactivity or sensitivity of bonds, or it could be used to break bonds. It has been noted that heating reactants in a microwave oven rather than over a Bunsen burner can speed up a reaction.

One useful trick in assessing the feasibility of a process is to see if it occurs in nature. There are at least half a dozen examples of the use of magnetic molecules in birds and insects, primarily to do with navigation or orientation. Plants almost certainly can use a magnetic field. Bracken can grow for a dozen metres underground in a dead straight line when seeking new fields to conquer, putting a shoot up every couple of metres or so to test the light values, which rather suggests an on-board compass. There are probably many more examples to be found, as little work has been done other than on iron bearing materials used for navigation.

At all events, there is a whole new field of chemistry to explore, and the early workers usually make the most striking gains.

MOBIUS AND MORE

If you bend a strip of paper into a loop and tape the ends together, you will have what could be a single link in a paper chain. It is a cylinder, having an inside and an outside surface, as can be shown by drawing a line around the long axis. If you start on the outside surface at the join, the line will be complete when you reach the join again, and the same can be done for the inside surface. If you push the loop flat, you will form two folds or reversals, of 180 degrees, of opposite sign. If you cut the loop longitudinally, it will fall into two separate loops of half the width of the original.

If however, before taping the ends together, you reverse one end by rotating it 180 degrees, you form a Mobius loop. This has only one surface, for if you start a line at the join, by the time you reach the join again, you will be on the opposite side of the strip and will have to go round again before completion. If you push the loop flat, you will get the two reverse folds as before, but the two sides will be crossed over, and the two folds cannot be at right angles to the strip for this reason. You will have to make another fold to get the strip flat. Three folds of 120 degrees to form a triangle is convenient; two are of one sign, and one the other. If you cut round the line, the strip will fall into a loop twice as long and half as wide as the starting loop. If you push this loop flat, you will find that the original twist of 180 degrees has been cut in half, and now occurs twice, so you will need two more folds to flatten it. If you cut this new longer loop longitudinally, you will find you have two loops, interlinked. Where they interlink, they are crossed over twice; that is, each loop is looped round the other, not just simply looped through as in a paper chain. Cutting one of these loops will give two more interlinked in the same manner. Putting more twists in the loop will give rise to a family of results, and a great tangle.

We can also make three dimensional Mobius loops. To do this, fold a strip of paper into three longitudinally, tape together into a

374

prism, and then rotate one third of a turn before taping the ends together. You can also rotate two thirds of a turn, a full turn and so on, developing a family of loops as before.

You can also fold the strip into four, making a square section bar, with again a number of fold options, and continue to make a five sided bar, and so on.

I will not spoil the fun, but leave you to experiment!

MAKING WAVES

If you look over the bow of a moving boat, you will see that the water is pushed up to form a bow wave; walk further aft and you will see that the water now forms the trough of the bow wave, as it falls back under gravity. This will repeat a number of times along the length of the boat. If you increase the speed of the boat, the wave will get bigger, and the wavelength will increase, so that there are not as many crests and troughs along the side of the boat. Eventually you will have only one wave with the boat sitting in the trough. Any increase in power will now not give much increase in speed, merely increasing the wave height.

From the air you will see that the bow wave forms a succession of parallel waves on either side at roughly forty five degrees to the boat, while a series of stern waves follows the boat progressing at the same speed in the direction of travel.

The stern wave can be useful if you are towing a dinghy; by adjusting the length of the towrope you will be able to position the dinghy so that it is sliding down the front of the stern wave, and the tension on the rope will approach zero. Let out the rope so that it is being pulled up the back of the stern wave and it will increase markedly.

This wave riding ability is used by planing vessels; a great deal of power is required to lift the boat over the bow wave. This is often called the 'hump speed'. The power requirement then drops markedly and the boat will start to slide over the surface of the water, with a reduction in wave-making and increase in speed. Surfboarders use the Severn Bore in the same way. Here the tidal wave is compressed into a narrowing channel and slowed, leading to a great increase in height; the oceanic tide is only a few inches. They can then ride the wave for about twenty miles.

In the days of canal traffic, fast passenger boats used to have a team of horses, like a stage coach, which dragged a light boat over

376

the top of the bow wave, which then acted like the Severn Bore. They could go as fast as a stage coach, faster quite often as they had a level surface, locks excepted. The postillion had a horn, to warn slower traffic, and the boat had a scythe at the bow to cut any ropes not dropped to let the faster boat through. I tried canal planing once; on the outskirts of Birmingham we found a length of canal with steel piling sides and no traffic. We had a forty horsepower engine on a 20 foot boat, which would do about 16 knots in open water. Opening the throttle pushed the bow up until it dropped quite suddenly and the boat was sliding down the bow wave, which was now effectively the stern wave. Speed was limited by the depth of the canal, but the throttle could be more than half closed. Counter-intuitively, for the bank-side observer, the first indication of the approaching boat was a drop in the water level and a flow towards the approaching boat.

Thus a canal boat could go faster than the same boat in open water, due to the containment of the wave between the canal banks. As the wave could not spread laterally, far less energy was required.

So why don't we build a canal to America? Actually, we don't need all of it, only the bits of bank that contain the bow wave, and we can take that along with us. Something of the sort is used in a catamaran; the bow wave is contained between the two hulls. Ideally the hulls should be slab sided, so no external bow wave is produced, but there is usually some departure from this. They effectively ride a soliton, a single wave; this can lead to problems. When they slow and come off the wave; the soliton can continue forward and catch other boats unaware. So if you are driving a fast catamaran, approach the harbour obliquely if you don't want to raise the ire of moored vessels within.

The method would be to 'plate' a boat; that is, put two metal plates either side of the hull, similar to a trimaran, but the two outboard hulls would have very little thickness. One would have to do experiments to find the best place and shortest length for the plates, and one would have to strut them together underneath to stiffen the whole. The skin resistance would increase, but this would be more than offset by the reduction in wave-making. The boat would weigh more, but engine size could be cut for the same speed, or alternatively cargo could be delivered faster. The additional side deck might be useful for unloading, or for extra bulky cargo where the cargo weight was light for the hold capacity. Early experiments would be with models, progressing to retro-fitting a small boat before proceeding to new-build.

IS THERE A DOCTOR IN THE

HOUSE?

Once, when at college, I sharpened a 4B pencil stub at both ends and then split the wood off to get two electrodes for an arc light. That did not work too well, so I taped my electrodes, connected to a dry cell, down on a microscope slide, and put a drop of pond water in between them.

I had read that fish migrate along an electric field; it is the potential gradient that counts, rather than current flow. It is used in conservancy to remove undesirable species like pike, or just to see what is there. It does not work well for deep sea fishing, due to the greater conductivity of the water, and longer distances, leading to a huge current demand. I had hoped my hydra might do backward somersaults (or maybe forward, it is difficult to tell if you are radially symmetrical), but they pulled in their tentacles and sulked. Some of the other little whizzy things seemed to congregate at one electrode however.

I did not publish; at that time I was enthused about private circulation papers. They still are very good if you want an idea taken up, but don't have the time or facilities to do it yourself. It is a bit like throwing money down in a school playground.

I did wonder at the time whether the technique could be used to get cancer cells to migrate, and in fact it is now used to encourage osteoblast migration to promote bone healing.

However, cancer cells can be very motile; they actively detach from a tumour and crawl about of their own volition. Some cancers are inoperable due to their site; brain cancers can be difficult to operate on because of the risk of collateral damage. However, it would be quite possible to put a hollow needle in, pass a small current, and see if they can be induced to migrate down the tube. A nutrient gradient could be introduced as well, with perhaps a growth factor at the distal end, and maybe a pressure differential. Even if only some of the cells were motile, those are the ones you want; better if they could be directed out rather than get loose in the bloodstream.

There could be mileage in the idea of a pressure differential too. While unselective, soft tissue can be sucked up a tube, particularly if an overpressure is used. One particular worry for hard hat divers is loss of pressure. If you suddenly decompress at a hundred and fifty feet, due to valve failure, you are likely to end up in the helmet! The Navy has done a lot of work on high pressure gases; they use two per cent of oxygen in helium at depths of a thousand feet or more, to prevent oxygen poisoning or nitrogen narcosis, the oxygen ratio being varied as to the depth.

I used to know a diver who set the world record at about fifteen hundred feet in a simulated dive in a pressure tank. That was thirty years ago, so they have probably gone further now. He said the main problem was boredom; you could get down in a few hours, but it took a fortnight to come back up, to avoid bubbles forming in the blood.

While one might not want to go to those extremes, it would be quite possible to use two or more atmospheres of pressure for operations where it was difficult to remove soft tissue. Instead of a scalpel we would have a vacuum cleaner.

And another thing; artemisin is the latest antimalarial. It may last some time, as it is dissimilar to the quinine derivatives, acting more like an oxidant. The Chinese found it by going through their old herbals. But did you know artemisia occurs in the British herbals too? For rheumatism, not malaria

It may be that the compound, or another associated one, is active against other conditions. It would be instructive to ask doctors using it to look out for this side effect. Drug companies of course look for side effects during testing, but not too hard, in case there are any. It would be a good idea to take testing away from the companies, and give it to a separate organisation for further investigation.

A FUTURE THEOLOGY

All major religions have a literature, and this falls naturally into two parts. The first deals with provenance. It is considered especially holy, being in the form of a revelation: God's message to man, or man's enlightenment regarding God. Examples are the Torah of Judaism, The Bible of the Christians, the Koran of Islam, the Bhagavad-Gita of the Hindus and so on. Although these are not the origins of the religion, they represent a nodal point and have come to have primary significance. Nowadays even the source material is being questioned. For example, the Bible of the Christians was cobbled together several hundred years after its origins, since when earlier material has turned up, which probably would have been included had it been around at the time. The Koran also has its problems, particularly as, if you read it in translation, it is not the Koran.

Some Christian material was excluded from the Bible, which on later interpretation might have been put in. Enoch for example was left out, and probably suppressed, perhaps because the angels in it were a trifle less than angelic, interbreeding with the locals. At all events, Queen Elizabeth I could not come by a copy, which she wanted probably because it mentioned a great southern continent. Even as late as Captain Cook's voyage to chart the other half of the coastline of Australia, he had sealed orders to look for it on route. There was probably one in the Vatican library, and one in Spain, but she was not on book-borrowing terms with either. Eventually one turned up in the eastern Mediterranean in Victorian times.

The second part is more to do with unwinding the detail; it is often not regarded as especially holy or immutable. Often doctrinal disputes lead to a split in the religion, as seen in the eastern and western branches of Christianity, later to split to the C of E and then innumerable oddments. In Islam, there are the Sunni and Shia groups, which are currently at war with each other. Other religions show the same pattern, and much of the literature is a subsequent interpretation of the primary material.

Comparative theology explores the similarities and differences between religions, giving an overview from a position of presumed neutrality, while archaeology and comparison of written evidence from outside the religion gives an historical setting and may support or question some of the statements embraced in the original material.

There is a third line of enquiry, at present largely unexplored, that could be taken up profitably. I propose to call this 'Operational Theology' until someone comes up with a better name. This is a development from the above, but with a different focus. The word operational is used to express the idea of an operation being performed. This could be compared to a mathematical operation. For example, we might divide the number one thousand and twenty one by seventeen. The focus is not on the thousand and twenty one; that is given, nor on the seventeen, for that is what we have decided to divide it by. We are primarily concerned with the answer, one hundred and thirteen, which is new information that we did not have before.

The concept is a method of evaluating alternates in theological thinking by studying their effects rather than comparing origins or ascribing validity by comparison to prescribed texts. The method is a general one that would have application to any ideas system, not necessarily religions. One is not concerned primarily with what the material is, or where it came from, but by what it does.

The outline of the enquiry is as follows.

Given an identifiable homologous group of people having a common belief system, or clearly defined set of beliefs or values, we may ask what effect this common belief has on:

1) Relationships between group members, as opposed to relationships between others, not in such a group.

2) Relationships between the group and other groups, or those not in any group.

3) Interaction between the group and the physical environment.

4) The effect of the group on the body politic.

Some brief examples can be given here. For example, many religions consider the giving of alms a mandatory or desirable trait in their members. This may take the form of giving a small coin to a beggar once a day. Its origins may have been that of supporting a holy man or seeker after enlightenment, with some reflected merit on the giver, rather than the relief of the indigenous poor, so its focus may have changed over the centuries. We can see how this plays out in the wider community.

Well, for one thing, it leads to an increase in beggars. Also, in some areas the beggars are organised coercively by gang masters, who transport them to their begging sites and take most of the money. Worse, there is an historic record of maiming young children to make them more effective beggars. To a lesser, or perhaps financially greater, extent the system still operates where the state has taken over the function of giving alms. Often it is in a person's interest not to seek their own advancement, for the benefits of the state are greater than those available in the early stages of providing for oneself. Thus an unintended self-fuelling poverty trap is created that may persist for generations.

Similarly, the Victorian reformers of the nineteen century in Britain abolished child labour in the cotton mills and other industries. The net result was to accelerate the shift of cotton weaving to India, where child labour was cheap or free, giving rise to a decline in the home industry. The problem had merely been exported. This is an early example of globalisation. Since then the trend has gathered pace, with the export of low pay and highly polluting industries to countries less able to resist them. The net result is the profit of the few at the expense of many.

The enquiry would be instructive; it might ease tensions between religions (or might increase them!). It might modify interpretation, and make the religion of more use to society. Currently religion is in decline in most countries. It was, in its day, a way of dealing with the problem of existence, and problems within the community. Whether we now have better ways is a moot point. The method could also be used in politics as well. Quite often political decisions are taken, and it is assumed that the problem has been fixed, without any tracking or evaluation to see if this is actually so.

At all events it should prove an interesting study.

A chap who worked for me went to sea after leaving school. He started well enough as an apprentice engineer on liners, but seemed to work his way down hill, ending his career on a five hundred ton coaster aground on a mud bank.

He could tell a good tale against himself however, which is a rare skill. His first ship was one of the then latest turbine liners; he had a yen to hear the turbine start up, the ship being alongside with the turbine decoupled for some work on the gearbox. (Not a gearbox like your car, but the speed of a turbine is far greater than that of the propeller, so some reduction is needed, and also a reverse.)

He cracked open the steam valve, but nothing happened, so he opened it some more. Again nothing, there being a lag while the stem pressure built in the pipes and condensation robbed the first pressure. So he gave it a good turn and the turbine gave a low 'oooh'; a bit more steam and it rose to an 'aaah', rising rather fast. He started to shut off steam while the turbine climbed to an 'eee', then an 'iii', tending to the edge of hearing. Luckily it slowly ran down without detonating, for a turbine running free can easily explode. There was a power station failure once when the turbine and anything near it were wrecked. The firm responsible took the unusual step of calling in other firms' expertise to solve the problem, which happened to be the wrong oil used in the valve-gear; it had carbonised and stuck the valves open.

If you travel by jet plane you will be comforted to know that there is a thick pipe of wrought steel surrounding the turbine, made in the same way as a gun barrel, to prevent any shrapnel penetrating the wings or cabin should it fail.

FATAL FLAWS

The escape velocity for the earth is often quoted as about seven miles per second. Actually you could escape at whatever velocity you like; what is really meant is the velocity a mass, falling from infinity, would reach on impact with the surface. For practical purposes infinity is redundant, as most of the acceleration occurs in the last few thousand miles. The strength of the field four thousand miles (one radius) above the earth is only one quarter of its value at the surface.

Newton would have been aware of this, for his theory of Universal Gravitation is a matter of the distance between masses, velocity, and gravity, obeying the inverse square law. It is possible that he examined what would happen if the mass continued its flight to the centre of the earth, as a mathematical exercise, for gravitating masses are normally dealt with as point sources located at the centre of the mass concerned.

It might seem a reasonable supposition that a finite mass, falling to the centre of another finite mass, would convert a finite amount of potential to kinetic energy.

This is not so, however; contra-intuitively, the amount of energy turns out to be infinite. The reason for this is that for every successive half radius fallen, the field increases by four times, due to the inverse square law, and the amount of energy converted doubles, as every successive half radius gives a fall of half the previous distance. It is quite possible that Newton was aware of this, but kept quiet about it, for those who did not accept his theory would have presented it as a fatal flaw. They had already raised the objection of 'action at a distance', which he could not give a mechanism for, merely stating that his results were those of observation.

In practice of course, for the real earth there would be a finite amount of energy available, for the acceleration would decrease once one was penetrating the denser material beneath the mantle. The gravitational field for a sphere is only that occasioned by the mass which lies beneath one's feet; all the rest merely cancels out. This is why galaxies rotate more nearly as a solid sheet, as opposed to the solar system where the inner planets rotate much faster; it all depends on the distribution of mass, radius and speed of rotation. There may be a sort of viscous drag levelling rotational speed as well, not so much because gravity is viscous, but because the gas and dust cloud is weakly so due to grazing impacts and shepherding. The same effect applies to globular clusters, gas clouds, and by extension to the universe as a whole. Whether a net zero field as at the centre of the earth or a black hole is the same thing as no field at all is a moot point; perhaps we could find out by flying a clock at a Legrange point.

It was over two hundred years later that a tentative proposal for a field of such strength as to be able to prevent the escape of light was proposed, not because light was slowed down, but because the wavelength increased to the point that it became flat. Whether there is anything left of the wave then is more a question of philosophy than science, for it would be undetectable.

The concept of a black hole solves some of the problems created by such a field, but gives rise to others, principally the infinite amount of energy created. We end up with a point of singularity, a macrocosm in a microcosm, a kind of infinitesimal infinity. As all stars rotate; the rate of rotation would increase as the radius decreased, partly because of the lesser circumference leading to an increase in spin rate, but also because all the kinetic energy created would be stored in the spin, the force being tangential for a spinning object. We can observe this for a spinning ice skater; as the outstretched arms are drawn in, work is done, and the rate of spin increases markedly in excess of that which would be occasioned by the redistribution of mass alone.

This immediately gives the problem of the centripetal force generated, which would also tend to infinity, preventing collapse to a point. Further, a transfer of mass would take place; energy is mass, and for a falling object the mass is increased by the equivalent of the kinetic energy imparted. This energy is provided by the inverse squares of the joint fields, and mass will be transferred from the greater to the lesser, tending towards, but never reaching, parity. Strong magnetic fields could further rob material from the centre to join the ring, or be ejected. Thus, as a star collapses, its gravitational field will provide energy to the infalling matter, giving mass transfer to the halo, which is prevented from reaching the centre by the centripetal effect of rotation. The final result will be a black hole loop rather than a point, the dimensions being determined by its initial conditions of mass and spin, and the energy value will not reach infinity. Thus a point singularity will never occur.

There is still a problem associated with the zero cross section of the string thus formed. At the centre of a gravitational mass there is a net zero field. With no mass outside of the net zero point or line, we are left with a kind of fossil field with no remaining abode or apparent means of support, rather similar to Berkeley's 'ghosts of departed quantities'. For a classical explanation this is perhaps another fatal flaw.

Quantum and relativistic considerations modify some of these problems; perhaps both space and time are quantised, or, if everything else is, it may not be meaningful to say they are not. If space is quantised we have a stacking problem. For example, the diagonal of a graph paper with quantum squares will be the same as the sum of the x and y axis, if one is not allowed to penetrate the quantum spaces, rather as an electron has to skirt the atoms in traversing a copper wire. Unfortunately this is scale invariant; your graph would be a kind of Mandelbrot staircase at whatever scale you looked at it, perhaps another fatal flaw. Other problems also arise. For example, if time stops at the event horizon then we have a field operating in zero time. This may be possible, but it would not be capable of doing work. Dispensing with field and supplanting it with bent space solves this problem but adds another dimension, which is likely to add complexity and potential for an infinite regress of problem solving if we are not careful. So far we have eleven dimensions; one is tempted to add "and counting".

Fatal flaws are of course no bad thing in an evolving theory; they are dead ends, which divert energy into finding other routes for explanation. Usually a theory is modified and quite often simplification occurs, the theory becoming more elegant and acceptable. Problems precede solutions, so perhaps we could invent a new paradigm for science as being primarily a problem seeking activity, rather than a problem solving one.

THIRD WORLD DEBT:

SOME PROPOSALS FOR A PARTIAL

SOLUTION

The problem of third world debt appears intractable. Some progress has been made, and promises, but the reality is that it is still rising and is now many trillions of dollars. Repayment, if it ever were on the cards, is likely to be negated by future events that have not been written into the scenario. There is a window of opportunity in the next ten to fifteen years or so however, that could be exploited, which will be examined in this paper.

The initial problem was caused by incorrect criteria for the loans. To avoid repetition of past mistakes, it is in order to detail them here.

For a borrower, the correct question to ask is 'will the deployment of the capital borrowed give rise to a return which is greater than the repayments of interest at an agreed rate, and repayments of capital over an agreed time?'

If these requirements are not met, the loan is unsafe. One could then ask 'what other sources of income or capital could be deployed to make up the anticipated shortfall in repayments, and, if available, is it actually worth propping up the loan in this manner?'

For the lender, the above conditions should also be examined, but normally there is a fall-back question: in the event of default, what assets could be seized to make good a deficit?

Retrospectively, it is obvious that these principles were not applied. There was confusion between the idea of loan and the idea of aid, there was an inherent and unjustified assumption that a country has an asset value, which it does not, for it cannot be repossessed or sold, and there was insufficient protection against fraud and mismanagement. Also money was not targeted for specific projects each with its own yield assessment. Adding unpayable interest to failed capital repayments then further exacerbated the problem. The results of these compounded errors are what we see today.

A partial solution to some of the debt is however possible. Debts are bought and sold, and bad debts are sold cheaply. Normally transfer of debt occurs as an adjustment between banks; this however leaves it still within the banking system, to give a net zero sum gain. It would be better if a debt could be sold outside the system; it would then be gone entirely off the books, at whatever loss, never to roll through the back door like a bad penny as a result of some other later adjustment.

The question is 'who would buy what is effectively worthless paper, at whatever price?'

The answer might lie with several of the various aid agencies, already working within the debtor nations. In other words, the debt could be internalised. The system would work like this:

The 'agency', and this might mean a blanket organisation acting as a feeder for several specific charities, missions, and aid and development agencies on the ground, would buy debt at deep discount, possibly starting with a (repayable) loan to get the process started. With the agreement of the 'host' government, this would be paid at face value in internal currency into an account in that country. This capital would then be deployed in specified target areas. There are many agencies and many priorities, but one fundamental priority would be to generate exports in order to recycle the debt; that is, to regenerate foreign capital in order to come back for a second slice, hopefully doubling up in the process. Once this priority was met, then excess funds could be deployed as aid in the normal manner.

The advantages of this system would be:

1) The banks would have reduced their debt burden both singly and collectively.

2) The debtor nation would similarly have reduced its external foreign currency debt burden, having internalised the debt.

3) This debt would then be spent internally to increase exports, which would further reduce the external debt, and to fund other activities in that area, which might best be described as 'crisis management'.

There are some provisos and safeguards that would need to be put in place. Firstly, exports thus generated should not supplant those already existing as part of the normal trade of the host country; new products should be sought, or an increase in current exports over and above the norm. If this is not put in place one might get merely a substitution, which would decrease the foreign capital coming in as a normal result of trade.

Secondly, viable product bases should be sought. Working with leverage, it would be very easy to start up new industry, only to see it fail as the real economy bit later and it had to pay its way without subsidy. Leverage would be acceptable for the start-up period however, as this would get things off the ground quickly and reduce the capital burden.

Thirdly, a watchful eye should be kept on any inflationary pressure. In other words, the injection of new capital in this way should be metered. There need not be an inflationary pressure in the long term. If one increased the money supply of a country by, say, one per cent by internalising some foreign debt, but also increased the goods and services available to be bought with that money by one per cent or more, there would be no inflationary pressure, the two being in balance.

Exports are not the only route; there is equivalence between exports and imports. For example, a country might not be able to produce complex plant and machinery, but could easily produce blackboard chalk, and a host of other simple low technology products. By exploring what is imported, and what could be made locally, substitution could be achieved, leading to a redeployment of foreign currency to more useful ends. 'An import saved is an export gained' should be the motto.

The principle to be observed is that sound business sense must be applied. There is a real debt to be reduced, with whatever leverage, and the net result must be better at the end than at the start, with viable industry left on the ground.

Earlier it was said that there was a window of opportunity, but that future events would overwhelm present projections of debt rescheduling, repayments and restructuring. Unfortunately there is another, unwelcome player on the scene. Current figures at mid-2001 for HIV infection in South Africa indicates over 19% of the population are infected. In Botswana the figure is 35%. Swaziland and Zimbabwe clock in at 25% with Zimbabwe and Zambia at 20%. Now we also have Ebola.

These are essentially yesterday's figures, for there is a delay between infection and testability, a delay in compiling and publishing figures, and a further delay before they are picked up and acted on. Also, it is not the total population that should be looked at, but the breeding population. Armed forces, and this may include fire and police services, tend to be tested more than the population at large, so sampling is more accurate. Here, the infection rate may be 50% to 60% in some African countries, with a possible if unbelievable 80% in Zimbabwe. The virus is of course still spreading, although more slowly, as it becomes more widespread and saturation is approached. Worse still, given a random association between an infected 50% male and an infected 50% female population, only 25% of couples will have both partners uninfected, reducing the viable birth rate still further. If infection is 80% the ratio drops to only 4% of children born to couples where both parents are uninfected.

Currently, about 40% of world births are in Africa, and it would seem there is no problem of depopulation. This is an illusion however, for the population growth rate is static, while the population death rate is increasing. Obviously the one will overtake the other eventually, so growth will slow, and then reverse. It is more a case of when than if.

There is a possibility of over-reporting, as this would attract more aid than, say, malaria, which is always with us. Under the present system there is little hope of a cure, for Big Pharma is more interested in an endemic treatable disease, which yields high profits, rather than a cure, which yields none. Considerable progress has been made in keeping people alive, but in many cases their output is less than the cost of treatment, and they are still ill. The progress towards cure for some other diseases is better. Malaria is a case in point; it is only with the intervention of large scale charitable funding that progress towards eradication is being made. Smallpox and polio are great success stories, which shows what can be done. Malaria was well on the way with DDT, but the mosquito became resistant in three separate areas. Artemisin is suffering the same fate, although administering three anti-malarials at once should give some breathing space.

We have grown used to populations doubling in 20 to 30 years in some countries. This is of course unsustainable; much of the useful land there ever was in Africa historically is now desert. True, the Sarhel is now greening, but this is perhaps partially the result of grazing stock being taken off as the people departed. Certainly the 'Granary of Rome', the North African shore, never recovered and the Saudi desert has lost game animals in historic times. The huge infrastructure cost by way of education, health and buildings is unachievable at this rate of increase, and many countries did not even manage to run on the same spot. Now we may have to get used to the idea of populations halving on a similar time-scale, particularly if Ebola or some other disease gets loose. Just how many successive halvings will occur is a matter for speculation and medical progress. Certainly much of the middle of the populations in some countries will be hollowed out. This will leave a ruck of elderly nearing the end of their lives, and orphans, many with HIV, to be cared for by a greatly reduced number of people who would have been in their most productive years economically. It has been said that banks are more dangerous than standing armies. Aids and Ebola are far more dangerous.

It might be a good idea for the worst or first affected countries to put in place an immigration policy, actively encouraging inward movement. Certainly the African country that does so would dominate the continent economically by the second half of the century, and become a powerhouse to drive development elsewhere. However, it might be difficult politically for recently ex colonial countries to open their borders to all and sundry, and by the time they do, there may be few or no takers. There are reputedly twenty million refugees in the world, but nearly all of them want to settle in the west. There may be some slight hope of 'reverse immigration' out of the USA, for a moderately skilled person there would become highly skilled by comparison in Africa, but perhaps few would go unless there were considerable cash incentives, which might alienate the indigenous population. However, there are some 'people of goodwill' who might be willing to undertake the task, and their effect would be disproportionate. There might be a case for an education program for survivors, for they will inherit a vastly changed world, with potentially great opportunities. The existence of such a program might itself encourage survival skills among the uninfected.

The concept of 'debt servicing' becomes ludicrous under these circumstances. There is little point in banks trying to collect the debts of the many dead from the few living. The requirement is for managed recession as numbers decline and productivity drops. There might be a temporary increase in agricultural products for export, before the unemployed population is used up and labour shortages reduce that too. The capital invested per head of population will increase, for there will be just as many factories, trucks and miles of road and railway as before, but these will decay over time and their value is problematic if there is not the population to work the plant. Unfortunately the per capita debt will increase also if something is not done about it.

The problem is compounded by the possibility of destabilisation. Some predictions embrace the collapse of the state in some countries. This need not necessarily be so, for the concept of 'The State' is not size sensitive; some quite small ones have existed for centuries. Some people think there may be a reversion to tribalism; at least that system worked for a long period. However, there will be internal and external pressures and a potential for conflict, for empty land is likely to be occupied, either by force or consent.

Quite what the end result will be is difficult to predict. A little should be said about the meaning of the term 'doubling time'. It means precisely that; i.e. the time taken for a doubling to occur. Bankers are familiar with the concept of compound interest. If money is invested at 7% for example, it will double in about ten years. Similarly a disease has a doubling time, which need not be constant, running faster through some populations or sections of a population than others; also, it will become longer as more people become infected. Once the 50% level has been exceeded, there can of course be no further doubling in that population. The point to be noted is that doubling can occur over a large number of years, without being noticed to any degree, until the final few doublings, when it becomes overwhelming, for each doubling increases the number achieved all previous doublings, plus one. For example HIV infection at one time had a doubling rate of a year, although it is slower now. After one year from the start two people would be infected, then four, eight, sixteen, thirty two, at the end of each successive year, and so on by arithmetic progression. Ten years on and we have roughly a thousand cases, an insignificant number. Another ten years of doubling and we have about a million, which is a significant number, although not yet a disaster. Another ten years and we have a billion, a figure we have not reached yet, but will do in a few years in the absence of other inputs. Death occurs several years after infection. We can call it seven years on average as a ball park figure, assuming the poor availability of drugs. If this is so, the death rate would peak seven years after the infection rate has peaked, and then decline in the same manner, similar to the half-life of a radioactive element. Currently we have around twenty million dead world-wide. This is more or less as predicted many years ago. If people started to die in the sixties, then this is the kind of figure we should have now, by repeat doubling. A simplistic extrapolation thus yields the infection rate overtaking the increase in population within a few doublings. This is unlikely to happen however, as there will be a major shift in perception and resource availability in the meantime. Smallpox was eliminated by a concerted effort, and some people have compared Aids to Smallpox, or to the Black Death, which was possibly Bubonic Plague or some variant thereof. There are some similarities; cellular infection may be by the same entry point as Black Death, and it was a world pandemic. However, about one per cent of the population was immune to Black Death, and about ten per cent recovered, while a large number of people did not get it all, probably because it acted faster, reducing the potential contact numbers for an infected person. Also there was not the population mobility then.

The best scenario is that Aids will become history in the next few decades and we will be worrying about the next emergent disease within this time frame. Hopefully there will be a medical input in the next decade, for the disease is certainly curable or preventable, if only we knew how to achieve this. The important thing about modelling is to know that the model predicts the future state of the model, not necessarily that of the real world.

The slow response time to the threat is surprising. This may be because threats of such a magnitude fall outside common experience, and are therefore not perceived in the same way as lesser threats. One failure is that a drug company cannot spend money on a disease until it becomes significant, therefore we perhaps should look more to public funding to give a 'first hard hit' for emergent diseases. We may not get as much breathing space next time. There are thousands of other viruses out there waiting, and a few might be worse than HIV; filoviruses, for example can kill in a few days. Ebola, a filovirus, unfortunately takes about three weeks, in which time one infected person can infect many others. Luckily fast diseases do not give as much time for infection of others as HIV does, so the spread can be more easily limited.

Another factor in the lack of initial response was misperception. HIV first gained public attention among the American West Coast homosexual community, and seemed to be largely confined there for a period. The problem was not homosexuality however but promiscuity, and the community at large did not understand the statistics of risk. There was an element of head in sand; first it was a homosexual problem, then an ethnic minority problem, then a kind of moral judgement problem, where the good were immune. Unfortunately the iatrogenic input was ignored; medicine as practised is not a science, being ruled in part by fashion and past behaviour, and is slow to change. Blood donations are collected individually, and there is no good reason other than convenience of handling why they should not remain so; however they are handled in bulk so that one infected sample can give a thousand new cases instead of one. The same mistake of bulk handling allowed the spread of CJD in cattle, and other diseases have been spread in this way.

The problem should not be considered a peculiarly African one. The disease burns faster there, probably because it originated there. In territories where there is, or has recently been a high infant mortality rate; there is also a societal pressure for a high birth rate, which is likely to give an accelerated infection rate. However, much the same scenario could be repeated in other countries, the West not excepted.

HIV is almost certainly the second biggest threat in the world, after global warming, for while we may talk of nuclear war, or a comet strike, these are of low or undetermined probability. HIV is a certainty. The only good thing is that a population reduction will reduce fossil fuel use, if not proportionately, for the most deaths are likely to be in low fuel use areas.

Against this rather sombre backdrop third world debt might appear almost irrelevant. However, the normal processes of trade, investment and commerce must go on. The window of opportunity merely becomes more urgent. Some real wealth could be generated out of paper debt, productivity per head increased as populations fall, research studies and hospices funded. Not all debt or all countries would be suitable, but in the absence of a total solution, we must pick away at those portions of the problem that appear most amenable. There is of course the same problem with the doubling time for debt relief; it will take time to get the program working. Given a starting capital of say ten million and a doubling time of a year to be able to come back with foreign exchange to buy twenty million of debt, the Aids problem will be racing ahead. Therefore time is of the essence, for the loss of one doubling time will halve the effectiveness of the program. What is needed is some form of rapid response unit, for even if we solve the problems of Third World Debt and HIV, and now Ebola, there will be other emergent problems in their wake.

We should perhaps examine the debt problem in a little more detail, for it is not only in the third world. Ireland, Portugal, Greece, Italy, the UK and America all have a debt problem. Even China is now in debt it seems, plus a lot of other countries. While it is easy to find who has debts, it is not so easy to find who 'owns' all this debt. How did they achieve this imbalance, and exactly what title do they have? The dominance in the market of immensely rich individuals and organisations is undesirable. One course to be pursued is the taxation of debt. If an individual has a lot of money, then they are relieved of some of it by taxation. If they own a lot of debt, then that also should be taxed. The receiver of the tax could then write it off or use it some other way. This would run down the debt over a period. The owners of the debt would lose nothing, as it is unrepayble anyway.

A property developer friend was called in to see his bank manager. The conversation went as follows. 'Are you aware that your interest payments are now over three thousand pounds a day? You should worry!'

To which my friend replied: 'You lent it to me. You worry!'

THE MPEMBA EFFECT

If equal quantities of water, in identical containers, one of which has previously cooled from boiling point, and the other just boiled, are exposed to freezing conditions, the one which has just boiled will freeze before the other. This was known in Greek times, and noted by Newton. It is called the Mpemba effect after an African schoolboy, who rediscovered it and pursued an enquiry with vigour. As this is probably an evaporative effect, this could be checked for first by comparing two sealed containers; if a difference in evaporation is the only cause, then they should both cool at the same time.

The question then is what is the difference between the two samples at the time equal temperatures are achieved?

Firstly, the boiled water will contain less air. One would expect however that the absorption of air would heat, not cool, the sample; as high velocity air molecules enter the water and become slower moving dissolved molecules, there will be a transfer of momentum, which should heat the water. However, as this occurs at the surface, it may be that the extra increment of energy provided by the incoming air removes molecules of water by evaporation, more heat thus being carried off than provided by the molecules arriving. Also, water has a surface tension, dependant on temperature, being near zero at boiling point, as there is no true surface due to molecules departing. Incoming air may also alter the value of this surface tension, as the surface is again less well defined. Experiments could be done with surfactants to see if they alter the effect.

An experiment could be done in which air free water and a normal sample are both exposed to freezing conditions, from the same start temperature. If air is a factor, the air free water should freeze faster, as in the case of boiled water. We could continue to compare freezing rate after the water has frozen; this would show if there were an effect not related to evaporation, or if evaporation was still accelerated from the ice surface, which might have a different structure.

However, water also has a 'structure'. Water is a polar molecule, meaning that the electric charge surrounding it is unbalanced, due to the hydrogen atoms forming a 'V' shape with the oxygen at the point. They therefore tend to form temporary pairs. These are easily broken by random thermal motion, but just as easily reform. As the water cools the random thermal energy declines and the pairs last longer, with the probability of other structures forming, such as hexagonal plates as in a snowflake nucleus. As these would form initially at the surface, where the inter-molecular forces are more unbalanced, a horizontal alignment of plates would be the most likely or preponderant arrangement. This would tend to inhibit vertical movement of random thermal molecules, and thus evaporative cooling. The structure would take time to form, so freshly boiled water would have less structure than water that had been standing for some time. It might be possible to detect layering by the refraction of light or sound in different samples of water, or by comparing electrical conductivity or movement of a dye in different planes.

An experiment could be performed in which two equal samples of unboiled water, one of which has been violently agitated, perhaps by forcing through a fine nozzle to form a mist or whisked in a blender, have their cooling rates compared. However, fine droplets might yield a greater number of platelets, if they exist, due to the large surface area created.

If a structure is present, and if it is destroyed by agitation, then the whisked sample should cool more quickly. A magnetic and/or electrostatic field could be used to see if this had any effect on molecular or platelet alignment.

It would be interesting to draw comparative graphs. One could be the cooling rate for freshly boiled water, another for 'ordinary' water, another for the cooling rate for water heated to 90 degrees C, but not boiled, then for water at 80 degrees, and so on. Similar graphs could be made for air free water and agitated water. A relaxation time could be established after which boiled water has the same characteristics as unboiled. It is quite possible that both effects are present and the observed effect is the net result of interplay between these, and possibly other effects.

An experiment to show that agitation alters the boiling point of water can be performed by placing a just boiled sample on a vibrating table. The water will boil quite strongly although below boiling point, if vibrated. This is not so much due to the energy added by vibration, but the breaking of bonds between molecules, the trigger increment carrying off more energy than supplied as explained before re air intake. In the same way, a fluctuating magnetic and/or electrostatic field could be applied to see if this had an effect on bonds.

We could see whether we can move from an effect, to a principle, from a one off example to a general class of phenomena. To do this we could perform the experiment with a variety of liquids, and see if any others give the same effect, and if polarity is a factor.

Since many industrial processes involve evaporation, reducing the boiling point to save energy would be useful. Similarly, we could do some experiments over the range to see which vibrational frequencies and amplitudes give the most energy efficient reduction in boiling temperature. These would probably vary as to temperature and class of liquid. Direction of vibration might also be a factor. The effect of a vibrational pulse travelling through a liquid near boiling point is to reduce the pressure so that it momentarily boils. By the time the elevated pressure part of the pulse arrives the bubble will have risen slightly, leaving the cooler water behind, so that it is not all recondensed as the pressure rises, thus leaving a bubble nucleus for the next pulse. This will be at a higher level, and therefore a lower pressure than before; also of course some of the bubbles will have escaped the surface and are thus not available for recondensation. Extremely high pressures can be set up by the collapse of a perfectly spherical bubble, so there may be an element of sudden radiative cooling as well.

There will also be some radiation loss from both samples; we could check to see if this is equal. It might be that water with a different structure radiates differently in different directions, thus we could compare the heat coming from the sides, bottom and top of the two containers. Mpemba's original experiment was with ice-cream, which has a different structure from pure water, depending on the mix, so it might be worth doing experiments with different solutions of gelatine, to see if the reduction in mobility for molecules as it sets alters the effect.

Convection cells will be set up on both samples, but the hotter will have stronger cells. These may persist after the equipoint is reached, perhaps because they are larger, stronger and faster. Similarly; there will be air convection cells above the surface. Fine dust in the water sample and smoke in the air could trace them. Also, it would be worth taking the temperature of the two samples at top, bottom and middle, to see if there is a greater stratification in the just boiled water.

Elimination of all the above potential effects may leave an unexplained nucleus to the process, which would provide material for further study, if present.

WALKING THE PLANCK

Let us suppose that you are a very small but intelligent insect, and you have been instructed to cross a sheet of graph paper to the opposite corner, following the shortest line, with the stipulation that you must keep to the lines. Initially you may go all along one side, and then along the other, which hardly seems the shortest distance. Next you set off going all along one side of the inch squares, then along the other, following a more diagonal path. This looks a bit better, but is the same distance; you then try dividing these into smaller and smaller squares, with the same result.

Unfortunately the sum of all the x-axis steps plus the y-axis steps still adds up to the same result.

Next you order the computer to generate some better graph paper. "Infinitely small squares!" you say. The computer does not do infinite however, but offers you the Planck distance as the next best thing. So you set off again, much slower this time; turning all those right angles with six feet is not as easy as you might suppose, to say nothing of keeping track in base six. However the distance travelled is still the same. Salami tactics don't work.

So we have a packing problem; exactly how do the squares in our Planck graph paper fit together? They might not even be squares any more, but a tessera of triangles, hexagons or anything else that will fit, so long as the sides are multiples of the Planck distance. We also have to consider the vertical stacking in the real 3D world.

Can we even have a diagonal? The square root of twice the square of a Planck distance comes out to a Planck distance and a bit, which might not be allowed. Exactly what is a bit of a quantum? It seems like a contradiction in terms, or for the logical, perhaps a terminal contradiction. However, if the diagonal of a Planck square has to be rounded down to one Planck distance, or rounded up to two, we have a funny kind of geometry. The trouble is, it should be scale invariant, and leak through into the real world, unless we have some kind of

401

quantum decoherence along the way. The first diagonal we can form is that of a three by four array, giving a diagonal of five. It seems we have an emergent property. You can make a list of subsequent diagonals if you like. Also, we have a problem with the time of transit for a signal; supposing that the length of time it takes for a signal to traverse a distance for one Plank is one time unit, which may also be quantised, the time for the diagonal would be the same, presumably. If not, we have a longer time for the diagonal transit, if time is not quantised in the same way. We also have a problem if we rotate our signal forty five degrees; it would take longer in one direction that the other. Maybe our squares are really circles, with nothing in between, in which case the x and y-axis, and the diagonal are all the same. We could even bend a Planck distance into a circle, in which case it would have no radius or diameter, so it would have nothing in the centre as well as nothing in between the circles.

Maybe we could label these non-existent dimensions 'unreal dimensions' similar to unreal numbers; combined together; this nested fiction might become a reality, a kind of double negative. These would not be the extra dimensions favoured by mathematicians in a tight spot, but something qualitatively different. A whole new geometry beckons. It might be rather abstract for the real world, but it could do for stacking universes in a multiverse.

Or we could have a vibrational state between one and two, possibly in another dimension, which is a quick fix for intractable problems, with the average coming out as the diagonal. This might explain why the vacuum is so busy, with quantum particles popping in and out of existence to take up the slack.

We can of course cheat by saying that Planck derived quantities are dimensionless, but this I think is a small step, or a quantum leap, too far. A dimensionless dimension is more of a subtraction than an abstraction, for the problem disappears along with any solution. It is rather like saying. 'I am not going to play any more'.

Euclidean geometry is much better. I can draw in a diagonal and measure it with a ruler, or even add squares and find the square root a la Pythagoras, who got his theorem from the Egyptians by the way; he was an undergraduate priest there for many years, and they taught him the easy bits they had known for a couple of thousand years.

Perhaps we could retain a conceptual Euclidean (or other) space, with items that inhabit it being quantised. Otherwise, if we divide it into Plank cubes it might freeze solid instead of having the zero viscosity we at present ascribe to it. A little bit of sub space between Planck elements might allow things to move, to introduce a little fluidity, for if there is nothing between quantum pairs, then there will be nothing but quantum pairs, however brief the dwell time. It might resolve the annoying discrepancy between observed and calculated quantum energy, currently the biggest miss-match around. It would also retain the mathematicians' infinitesimals. Come to think of it, there should be some very small element of viscosity caused by quantum pairs. If they have mass, however fleetingly, they should impede the progress of fast particles over long distances. Instantaneous events don't work, for if an object is instantaneously there it is also instantaneously not there and can be discarded. There must be a ratio between time there and time not there.

That's one good thing about geometry; analogue works!

THE MIND AS AN ENTROPIC

SYSTEM

When I was four I was taken to see *Snow White*, the first and last performance in our town, for it was 1940 and the cinema was firebombed that evening. Everybody smoked in cinemas in those days, and I was fascinated by all those little moving lights from the projector, like little searchlights or torch beams through the smoky atmosphere, familiar to me, for there were no street lights then. They fanned out and moved, changing colour and intensity and eventually welding seamlessly into a moving picture on a screen. I thought there must be a most marvellous machine to do all that. When I found later that it was achieved by shining a light through a celluloid strip, I felt cheated.

Later, in my early teens, in an RE debate, I found myself defending the resurrection of the dead: not that I believed in it, but you need two sides in a debate. I am not at all sure that the visiting clergyman who took our lessons believed in it either, but he believed at it, which is the next best thing.

My argument was that the body is a finite system, composed of a few trillion atoms, arranged in a particular manner. If by some marvellous machine, we could form such an assemblage, we could reconstruct the body. Therefore there was a theoretical, if impractical, possibility of resurrection. The first time round was the impossible thing, a repeat merely difficult.

Now of course we have cloning from a single adult cell. Cheated again! The debate is essentially over. If you want a new body, you can have one. Your mind won't be in its head, of course, so there is no great advantage, but considering that we are only in the first second of the first day of this particular venture, so to speak, we cannot draw a line under that yet.

In my twenties, as now, the debate was about intelligent machines.

If only we could refine our computers, patterning them on the human brain, we could reconstruct a mind and live in the computer, another most marvellous machine. Actually, the debate was there in the sixteen hundreds; automatons then could sign a name, speak a few words, play a tune, and dance a jig. If only we could refine our clockwork.

I am not going to be caught out a third time. If we ever do any such thing it will probably be as simple as 'copy to floppy' on your home computer; not that we do that anymore, but it rhymes, and a body without a mind might be a bit floppy.

Which brings me to the problem of the mind, or 'soul' as it is sometimes still called. There is a persistent but scientifically unjustified belief that it can survive the death of the body. To be accurate, the theological question is, 'Is there life before death?' which is a slightly different perception, but be that as it may. The belief is very old; the quantitative and more pertinently qualitative, survival of some kind of essence is the core of most religions. The Egyptians even embalmed their dead Pharaohs to prevent reincarnation, so that they stayed with the gods, looking after Egypt. The modern position is of course that it goes nowhere, and the explanation is that the belief is fostered by a desire for survival. I disagree with this explanation. Most people do not want to live forever, and don't want to be particularly good: well, no more than the average anyway, as is evinced by their lifestyle. Even the Jews did not believe in reincarnation before they went into Egypt. Most of our religion comes from that root, for they went in with little and came out with much.

Also, I have checked the address, and surprisingly, nowhere is not a destination. It does not exist, and therefore you cannot consign things there.

The reason is to do with entropy. The universe is a thermodynamic system that is cooling and 'running down'. The process is directional and irreversible. Thompson put it in a nutshell as recently as the nineteen twenties, with the 'Heat death of the Universe', although the Greeks knew it in general terms thousands of years before; it will end up cold and old. We all believed this with all the enthusiasm and unquestioning acceptance accorded to the statements of our scientific elders and betters, for a while that is.

There is however the less quoted third law of thermodynamics, which states that as the temperature drops, order increases. We can see it in crystallisation, and we can see it in the periodic table, for the elements are more complex than the primordial hydrogen and helium they came from. At another level, a single cell from a blade of grass in your lawn is more complex than the sun, which gave it life. The universe is not so much decaying to entropy, as decaying to order, or information. The conclusion is not easily disputable, for the maths is the same. Time was when we dumped nuclear waste in the sea, and thought it would go away. It didn't, and neither will entropy. We cannot just consign it to the dustbin, for we will need a bigger dustbin to put that one in. The logical derivation of this is that information may be indestructible, for it is entropic. Thus the trend to order, information, and perhaps mind may be a natural and inevitable process, rather than a special case.

There is also the possibility that the 'decay to information' may only be sequential from our viewpoint; it may be concomitant all through the process or even a necessary precursor to it. Democritus saw only 'atoms and the void'; the atoms have come to pieces somewhat since then. Now we see little packets of energy, and ponderable matter has been discarded, while empty space has become a little more complex. Take one step further and we may be left with ponderable information. John said it first; 'In the beginning was the Information, and the Information was with the Universe, and the Information was the Universe.' We can lose information, of course. I do it all the time, or more accurately, I mislay it, not destroy it. Even if we lose the civilisation, or perhaps when, the periodic table will still be in the rocks, waiting to be read by anyone who comes later.

The problem is further compounded by the fact that information is a zero mass phenomenon, a matter of quality, rather than a quantity of mass. You could, for example, take your copy of this article, slice it up into individual letters and give them a good shake up in a bag. Trivially, you are likely to lose some dust, and cut cellulose bonds may well form an association with water or oxygen, but you would not expect the weight to change because of the randomising of the letters. If you believe this might be so, get a second experimenter who does not understand English to check your weighing. You would not expect different results according to the prior understanding of the material. Actually, you will not really have destroyed the information by cutting up one article, so perhaps the argument lacks weight.

Zero mass phenomena are often achronous and non-local, that is, not located in any particular patch of time or space. Ride a photon and you have instant elsewhere; there is no time of flight so far as it is concerned, no distance either, for that matter. Ask a photon the time and it will tell you it is now; and now is forever. For the photon, the universe is flat. Neither was there any big bang; the twin concepts of time and distance are not in the vocabulary of the photon. Their beginnings are their ends. They shall not grow old, as we that are hadrons shall grow old. The information processing in your brain, or your computer, is essentially based on photon exchange; not visible light, but electromagnetic quanta of various wavelengths. Photons are the way atoms or electrons communicate with each other. If thermodynamics is the most all-embracing theory we have, to the extent that all other theories are subsets of it, then quantum electrodynamics is held to be the most perfect scientific theory we have, one of our twin pillars of the known world. If it laps over into the way we think, then so be it.

The mind can therefore be considered as an interactive information system of zero mass, which may not necessarily be as time-locked as the body.

I have no axe to grind, but I think the obituary of the mind, as being some mechanistic system, is premature. There is something alive in your head and it is not just brain cells. They are the hardware, you, the software running. There is another hierarchy to be explored, a qualitative leap just as great as the one from inert matter to living organisms. Better to admit that both sides of the debate are as yet in an area of profound ignorance. Sticking one's head in the dustbin and saying the problem does not exist will not serve, for one still has to get rid of the dustbin. We have a disposal problem. So, can the soul live forever? I am sure I don't know, and nor does anyone else either.

I don't state this as a position, although I am sure some will support it and others not, but merely as a comment on the present state of the debate. The universe does seem to be decaying to entropy and this can be construed as decay to information. Perhaps it is in some way becoming self-aware, and we are part of this process. It is unlikely that we are the front runner in this, as there are older stars out there. So why haven't they communicated? Maybe they have, but not by radio. That is far too slow. Perhaps there is an entry level to the debate. What are the qualifications? Why, to be able to communicate, of course!

CONCLUSION

I hope you have enjoyed reading this book, and are prompted to do a bit of inventing of your own. As the book is largely unadvertised, I hope you will also tell your friends that it is a 'Must Have', and if they make a million as a result of your introduction, they should put a bit your way; after all, one good turn deserves another!

It has been said that there are three ways of making money: Wine, Women and Inventions, and that Inventions are by far the most lucrative of the three. Unfortunately, it is seldom the inventor that reaps the benefit, but you need not tell them that.

I put here some addresses that may be of use:

Intellectual Property Office. (Previously the Patent Office)

Concept House, Cardiff Rd, Newport. S. Wales, NP10 8QQ.

www.ipo.gov.uk

They have a number of useful publications, and a provisional patent for one year is free.

The Institute of Patentees and Inventors

PO Box 39296 London, SE3 7WH. UK.

https://www.ipo.gov.uk

At the time of writing there is talk of a link with Salford University, which looks interesting. They will advise on various aspects of invention, and they hold meetings and exhibitions where you can meet others of like bent.

There may be a local inventors club near you; scan the local directories for details.

British Mensa Ltd. St. John's House, St John's Square,

Wolverhampton, WV2 4AH. UK.

http://www.mensa.org

The International Mensa website gives links to branches in many countries. You can go to a meeting without being a member, if invited by a friend, or the local secretary would probably take you along. You might not notice that they are particularly intelligent, but you are likely to meet a greater diversity of interests than in the normal run of things, and some interesting people. Join if you like; if someone is interested enough to take the I.Q. test, they are well on the way. It can be very useful when travelling, for it is an international society, so don't be put off. For the rest, I have heard there is an alternative society called Densa, but I do not have an address.

The World Future Society.7910 Woodmont Ave., Suite 450 Bethesda, Maryland 20814. USA.

http://www.wfs.org

The Magazine will keep you abreast of current thinking on the subject, again international, but mostly in America. As a member you can attend meetings, and again make a circle of contacts that would be difficult to achieve otherwise.

'New Scientist' and 'Scientific American' magazines are always a good read, with 'Nature' for more detail.

ABOUT THE AUTHOR

Alan Thompson was born in Hampshire in 1936. After twenty years spent teaching science in England, Kenya and Saudi Arabia, he moved to the Isle of Wight to run a family holiday business with his wife and two children, in a large and rather rambling Victorian mansion. Now retired, he lives in a converted chapel in Pilley, a New Forest village, where he is still active in writing and invention. He is a member of the Institute of Patentees & Inventors, the World Future Society and MENSA. His interests are invention, sailing and botany.

He has written many articles for several magazines, including 'The Inventor,' 'The Futurist', and 'Mensa Magazine', plus some Science Fiction short stories for New English.Library, and one book, 'Understanding Futurology,' published by David and Charles. English edition ISBN 0-7153-7761-2. Portugese edition ISBN 85-220-0025-5.

Contact : atwindshome@ yahoo.co.uk

www.ingramcontent.com/pod-product-compliance
Lightning Source LLC
Chambersburg PA
CBHW051849170526
45168CB00001B/35